HOW TO USE A
MICROSCOPE

W. G. HARTLEY received his Bachelor of Science degree from London University and is by profession a fresh-water fisheries biologist specializing in the Atlantic salmon. In addition to and in conjunction with this work, he has been particularly concerned with the instrumentation and evolution of design of the microscope. In 1946, he constructed a variable phase-contrast microscope, later exhibited at the Oxford International Symposium of the Physiological Society and at the Exhibition of the Physical Society; it is believed to be one of the first of its kind. A Fellow of the Royal Microscopical Society for the past twenty years and a member of the Quekett Microscopical Club since 1950, W. G. Hartley is also the author of numerous articles and papers on microscopical and fisheries matters.

BERNARD FRIEDMAN is Technical Director of Scientific Instruments at Nikon, Inc., a manufacturer of microscopes and other optical instruments. A member of the Royal Microscopical Society and of the Board of Managers of the New York Microscopical Society, he previously taught biology in the New York City school system and presently lectures and teaches classes on microscopy throughout the country.

JOHN J. LEE is Director of the Living Foraminifera Laboratory of the Micropaleontology Department at The American Museum of Natural History and Assistant Professor of Biology at New York University. His research on the fine structure and reproduction of various protozoa and algae has given him wide experience in microscopy and recording microscopic images. He has also designed special attachments for advanced microscopic research.

W. G. HARTLEY received his Bachelor of Science degree from London University and is by profession a fresh-water fisheries biologist specializing in the Atlantic salmon. In addition to and in connection with this work he has been particularly concerned with the instrumentation and evolution of design of the microscope. In 1946, he constructed a variable phase-contrast microscope, later exhibited at the Oxford International Symposium of the Physiological Society and at the Exhibition of the Physical Society; it is believed to be one of the first of its kind. A Fellow of the Royal Microscopical Society for the past twenty years and a member of the Quekett Microscopical Club since 1936, W. G. Hartley is also the author of numerous articles and papers on microscopical and fisheries matters.

BERNARD BAUMAN is Technical Director of Scientific Instruments aktuell, Inc., a manufacturer of microscopes and other optical instruments. A member of the Royal Microscopical Society and of the Board of Managers of the New York Microscopical Society, he previously taught biology in the New York City school system and presently teaches and conducts classes on microscopy throughout the country.

JOHN J. LEE is Director of the Living Foraminifera Laboratory of the Micropaleontology Department at The American Museum of Natural History and Assistant Professor of Biology at New York University. His research on the fine structure and reproduction of various protozoa and algae has given him wide experience in microscopy and recording microscopic images. He has also designed special attachments for advanced microscopic research.

HOW TO USE A MICROSCOPE

W. G. Hartley

AMERICAN EDITION EDITED BY
JOHN J. LEE AND BERNARD FRIEDMAN

AMERICAN MUSEUM SCIENCE BOOKS
Published for
The American Museum of Natural History

The Natural History Press
Garden City, New York
1964

The line illustrations for this book were prepared by the Graphic Arts Division of The American Museum of Natural History.

How to Use a Microscope was originally published in England under the title *Microscopy* by the English Universities Press Ltd., in 1962. The American Museum Science Books edition is published by arrangement with the English Universities Press Ltd.

American Museum Science Books edition: 1964

Library of Congress Catalog Card Number 64–20578

Printed in the United States of America

Dedicated to

J. W. KETTLEWELL, B.A.

In recognition of his services to
the Royal Microscopical Society

CONTENTS

FOREWORD

Microscopy resembles geography in forming the
ground common to half a dozen sciences which are
otherwise unrelated. For this reason none of the many
books already dealing with the subject represents the
sum total of what every microscopist ought to know.
No author could deal with every possible contin-
gency from personal experience, and an encyclopedia
would be out of date long before publication; one
author writes admirably for students of optics, one
for biologists, and others for chemists and geologists,
and books at varying levels of authority cater for
other audiences.

There remains the increasing number of serious
students who are not mathematically inclined, and
who may lack a training in physics, but who need
to use the microscope in the course of their studies.
It is customary to pay them the compliment of assuming that at the instant they first grasp a microscope they automatically become endowed with the
ability to use it effectively. Consequently, although
details of specimen preparation are carefully explained, the examination of the specimen and its interpretation are usually left to common sense, and
the apparatus repaired when necessary.

Experience suggests that these students would find
work easier if they had sufficient background information to know the appropriate ways of examining
specimens of various kinds, and, above all, what
their appearances signify. Some of them will be go-

ing on to do responsible work, possibly with complex
electronic instruments, and if they already appreciate
which features of the microscope image are due to
the conditions of its examination a great deal of future confusion can be eliminated. The purpose of
this book is to supply all the necessary basic information about the microscope as an instrument and
to provide the essential background knowledge for
its use.

                                        W. G. HARTLEY

PREFACE TO THE AMERICAN EDITION

Up until now, books available on microscopy have been either too detailed for the beginner or too skimpy for those who develop more than a passing interest in the microscope. For many years, there has been a need for a book on the microscope written for the serious beginner—a book devoted to the instrument itself rather than specimen preparation. This was W. G. Hartley's avowed intention, and we present his book to readers in this country with the firm conviction that he has succeeded admirably.

The optical considerations involved in microscope image promotion are amply illustrated and developed lucidly without getting into involved mathematical optics. At the same time, the dangers of oversimplification are for the most part avoided and the necessary ground is covered quite thoroughly. The discussion of image interpretation is particularly important and effective. The methods of securing optimum image quality by the proper adjustment of every component from light source to eyepiece are described with practicality, along with discussion of the theoretical significance of each step. There is, in addition, an introduction to special methods such as phase contrast and interference microscopy, and photomicrography.

With this gem at hand, the editors have had to make only minor revisions of the English edition, such as bringing the terminology more into line with domestic usage and referring to makes and components in more general use in the United States than in Great Britain. The section on photomicrography has been enlarged and revised to include recent developments in techniques. The bibliography has been enlarged, a glossary has been added, and subheadings have been inserted for improved organization of various parts of the book.

The book, however, is Hartley's, and the full credit for its creation belongs to him.

New York City   JOHN J. LEE
March 1964   BERNARD FRIEDMAN

HOW TO USE A
MICROSCOPE

Chapter One

GENERAL CONSIDERATIONS

Definition of Microscopy

Microscopy has been defined with admirable simplicity as the use of the microscope. Unfortunately this definition is not entirely lucid unless the microscope is also defined, and it is very difficult to do this in a way which will be universally acceptable.

Practically everybody can recognize the appearance of a conventional microscope, which is accordingly included, as a kind of hall-mark, in every picture of any kind of scientist, usually in surprising company with a chemical retort, and is also depicted to suggest painstaking and abstract high-mindedness in advertisements.

In actual fact, quite apart from its applications in research and education, the microscope is an everyday tool in the majority of manufacturing processes. The ordinary man is practically enveloped in the results of microscopical work, from the leather soles of his shoes to the crown of his felt hat. As a result of this wide employment, the instrument assumes different appearances, being specialized in one way or another; one user may be able to regard it as a superior kind of reading glass, whilst to another it is essentially an analytical instrument.

The difficulty of a good descriptive definition has been increased by the fact that the microscope and its use have entered a phase of rapid development after almost a century of relatively imperceptible refinement, which had seemed to be approaching a final limit. These new developments result from unexpected advances in optical theory, and design and manufacture have improved with experience of mass production. At the same time rapid advances in biochemistry have provided the microscopist with techniques of almost incredible exactitude, whilst the application in industry of refined physical methods of measurement has provided the tools for the development of correspondingly more sensitive physical methods in the laboratory.

Unfortunately the reaction of the microscope users in the face of these windfalls has been to concentrate on the treatment of the specimen, and to accept a somewhat mediocre standard of instrument manipulation and choice. Whilst it may be accepted that the use of the microscope is a means to an end, it still remains true that the successful use of any kind of instrument requires some appreciation of what it is intended to do, and the limitations inherent in its design and construction. If these aspects are not taken into consideration, not only is the work itself rendered more difficult, but the results obtained are questionable. It may not be appreciated that there is any possibility of obtaining misleading results when something is "merely magnified", but mistakes are easily made; crystalline substances are confused, organisms are overlooked, or features are described falsely and confuse subsequent workers.

Visibility and Resolution

The microscope is an instrument for rendering fine detail visible. It follows from this definition, (which describes the function of the instrument, not the microscope itself), that three factors are essential to its working: an object to be observed, light to observe it by, and an eye to observe it with. Any result obtained depends on the interaction of these three, which all have their peculiarities and limitations. It is the province of the microscopist to take account of these, and to exploit them to the best advantage.

Anything which is to be seen must conform to two essential requirements; it must be intrinsically visible, and it must be large enough to be seen. These conditions are quite distinct, but tend to be confused together. An object is visible because it affects in some way the light which reaches the eye from it. A speck of dust in a sunbeam is visible because it scatters light, and appears luminous, though it may be far too small to be perceived in any other circumstances, and the same is true of the stars, which are conventionally taken in optical examples to represent luminous points without size. Conversely, although no one could doubt the existence of localized variations in the air on a windy day, these pressure differences are not visible except by their physical effects, although they may be very extensive.

The concept of **visibility** may be extended to embrace conditions in which a camera is recording the action of a specimen on infra-red rays, or a television screen the results of electron bombardment. These resemble vision in that in each case a suitable stimulus produces local effects on a sensitive surface, and this is ultimately examined by the eye of the observer. If an inappropriate stimulus is applied, the receiving surface will not respond to it; the eye cannot detect ultra-violet or infra-red images because it is insensitive at these wavelengths, and the ordinary photographic plate is indifferent to colour. Neither the eye nor the camera can recognize an image consisting only of a pattern of phase differences until it is translated by some optical trick into an intensity pattern, which is visible.

Occasionally the reverse condition may be encountered; a well-defined image can be seen, photographed, and measured, but proves to have no connection with the specimen, which can be moved or even changed without affecting the appearance. This is rarely due to defective apparatus, but usually to defective microscopy—the user is doing something he does not appreciate. The laws of optics are quite impersonal, and always answer a fool according to his folly.

It will be necessary later to discuss in detail the various properties which control visibility. At present it is sufficient to draw attention to the fact that the production of a visible image is a deliberate act, and commonly provides the microscopist with his most exacting task.

The impossibility of observing things which are too small is familiar with ordinary experience. The human eye can separate with little difficulty lines as close as one hundred to the inch at ten inches distance from the eye, and in ideal circumstances a **resolving power** of four times this value can be attained, corresponding to the appreciation of an angle equal to one minute of arc.

The Eye

Essentially, the eye consists (Fig. 1) of a **lens,**

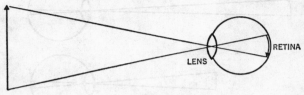

Fig. 1

which forms an image of the scene before it on the **retina,** a mosaic screen of specialized cells which are sensitive to light. The sensitive cells are very small and very closely packed, and in ordinary circumstances provide all the information necessary about any object sufficiently prominent in the field of view to excite interest.

There must obviously be a limit to the concentration of detail which the retina can accept, and this is reached when the individual details of the image are too close together to fall on separate receptors; the individual receptor can react to a greater or lesser stimulus, but it cannot discriminate between simultaneous stimuli. It follows that a scene or object can only be perceived as minutely as the structure of the eye permits; if a finer examination is necessary, the image on the retina must be spread out until the components become distinct. In other words, a portion of the field of view must be made to cover the entire retina.

In ordinary circumstances this is accomplished by

taking "a closer look" at the object, holding it closer to the eye, so that it occupies more of the field of view (Fig. 2). There is a limit to this expedient, how-

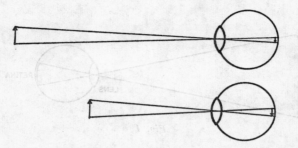

Fig. 2. Size of retinal image increased by bringing object closer to eye.

ever, as the automatic focusing of the normal human eye is restricted to distances of ten inches and longer. A still closer look can only be simulated by spreading out the image on the retina artificially (Fig. 3). This spreading out is the function of the magnification provided by the microscope.

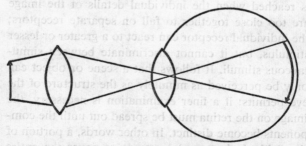

Fig. 3. A magnifying glass enables the eye to image objects very close to it.

Limitations of the Image

The production of an image of any selected size presents no great difficulty; it may appear paradoxical, but it is nevertheless true, that the magnification of a microscope is its least important attribute. The essential feature is the retention of the desired detail in the magnified image.

Ideally, the system forming the image should collect every ray of light leaving the object, transmit it to the position of the image, and there re-constitute the pattern of light waves characteristic of the object. To the extent that it fails to do so, by wasting the information originally available or reassembling it faultily, it will cause the image to differ from the object.

Perfection cannot be attained in the microscope, as much of the light from the object fails to enter the instrument in the first case, (Fig. 4) though such light as is collected can be brought to the image in a very accurate approximation to its original distribu-

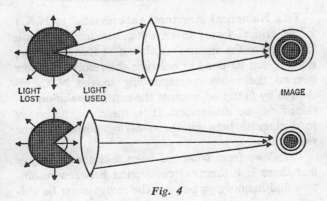

LIGHT LOST   LIGHT USED   IMAGE

Fig. 4

tion in the object, and this represents a technical triumph of no mean order, as will be explained later, which can only be achieved in definite circumstances.

The image can never, therefore, be quite as distinct as the original object, and the nature of light itself provides a limit to the perfection attainable, as rays leaving a single point in the object can never be recombined to form another true point, but only a minute disc instead. Consequently a limit to detail rendering is reached when the discs representing adjacent points in the object become confluent.

Bearing these facts in mind, the performance of a lens is a mathematical consequence which can be calculated on paper and checked in practice to confirm that the results are in agreement. The most important characteristic to emerge from these calculations is that the ability to separate fine detail, known as the **resolving power,** is directly proportional to the light-grasp of the microscope, which in mathematical form is designated the **Numerical Aperture.**

Necessary and Excessive Magnification

This Numerical Aperture, (abbreviated to **N.A.**) is the critical factor in an optical system. It sets a final limit to the amount which can be seen by its use; nothing can be done to enhance the resolving power beyond the value corresponding to the N.A., although by faulty adjustment the actual **resolution** obtained can be diminished. It is, therefore, necessary to distinguish between the **resolving power** and the **resolution.**

It follows from what has been said about the eye that there is a practical connection between resolution and magnifying power. The image must be sufficiently large to spread the details over the cells of the retina widely enough for them to be appreciated, but if all the detail a lens will reveal is magnified sufficiently for clear discrimination, but no more, the best result is obtained. Further magnification does not provide any new details, and tends to make those already present less distinct; the image becomes "woolly" as the existing features are spread out to fill the extended area (Fig. 5).

It might be thought that the obvious course would be always to employ the maximum resolving power, but this is far from being the case. The higher the power of a lens, the more restricted its application; it reveals more and more of less and less, and it is a sound principle always to use the minimum power which will reveal what is required—it is easier, quicker, and less expensive.

If the lens will not magnify sufficiently to separate

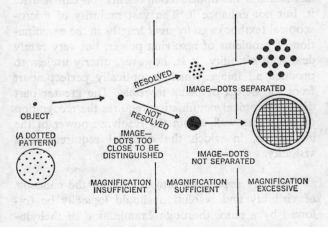

Fig. 5. The relationship between resolution and necessary magnification.

the details which it can resolve, it is inefficient if used for direct visual observation, though it is efficient if used to take a photograph for subsequent enlargement which will enable the hidden details to be displayed. The resolving power of a photographic plate is greater than that of the eye, and this is the principal method of obtaining the utmost resolution.

Owing to the practical connection between resolution and necessary magnification in visual use, microscope lenses are constructed with a magnification which is greater in those with a higher resolving power, thus avoiding unnecessary difficulty in design and use, and providing for the widest possible range of application.

It will be noticed that as the resolving power is controlled by the nature of the eye and the construction of the lens system, the microscopist has practically exhausted his voluntary control of it when he has selected his optical components; he can restrict it, but not enhance it. The vast majority of microscopical textbooks go to great lengths in the examination of problems of resolving power, but very rarely deal with visibility. It is, however, utterly useless to produce an image which is optically perfect apart from the defect of being invisible. The greater part of microscopical examination requires the recognition of something well within the resolving power of the instrument, in which the essential requirement is visibility.

This very superficial introduction of the concepts of visibility and resolution should logically be followed by a more thorough examination of their influence on microscope manipulation, but to make this intelligible it is essential first to introduce the ordinary handling of a microscope, after which the ways of obtaining the best results will become self-evident.

From what has already been explained, it will be obvious that the manipulation of the microscope must be directed towards making the object examined provide a picture which is clearly visible, and sufficiently large for convenient examination.

A clearly visible picture is one which is well-defined, adequately illuminated without being embarrassingly brilliant, and free from the deceptive effects of shadows, excessive contrast or uneven lighting. Haziness or obstructions to free observation are detrimental, because the former diminishes the definition of the picture and the latter interferes with its appreciation.

The microscope is designed to provide these conditions, and the user has the option of producing exceptional lighting effects for use in special circumstances. As a rule, the picture is produced partly by the illumination applied to the specimen, and partly by preliminary treatment of the specimen to ensure that it behaves in a definite manner. Both of these factors are important, and they are complementary. In some cases it is impossible to perform any preparative operation on the specimen, and then the results attainable are necessarily incomplete, though they may be adequate. More usually elaborate specimen preparation is followed by elementary methods of examination. This also produces incomplete results, though again, the results required may be achieved; the danger is that the user fails to realize that ambiguities which he recognizes but cannot resolve are often capable of being elucidated by quite simple optical means.

Chapter Two

THE MICROSCOPE—DEVELOPMENT AND CONSTRUCTION

History of the Present Design

Although recent technical developments have led to considerable variations in its shape, the appearance of the conventional microscope is familiar enough. Usually it consists of a metal stand supporting groups of lenses, and is, therefore, traditionally divided into "brass" and "glass" by microscopists. The glass parts —lenses, mirrors, prisms, filters—are the essential components which enable the specimen to be examined, and the brass parts are the necessary mechanical auxiliaries which enable the glass parts to be adjusted.

The basis of the design as a whole is most easily followed in an historical sequence explaining the transformation of the reading glass into a compound microscope; the following thumb-nail sketch covers the main course of innovation.

1. It was known, certainly in the Middle Ages and probably earlier, that a convex lens would provide a magnified image.

2. About 1600 it was discovered at Middleburg in Holland that two such lenses could be mounted at the ends of a tube, and would then produce a

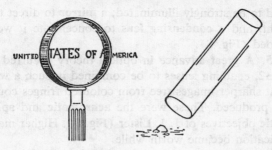

UNITED STATES OF AMERICA

Fig. 6

greatly magnified image of an object held in front of the tube. (This is the origin of the **objective,** the **body,** and the **eyepiece.**) By comparison with a single lens of equal magnifying power, the compound instrument was more convenient in use, though the single lens gave a better image.

3. It was necessary to locate the specimen precisely with respect to the body to keep it in view; this leads to the **stage** and the **focusing movements,** and also to a supporting **arm** and **base.**

4. The enlarged image was so dim that the object

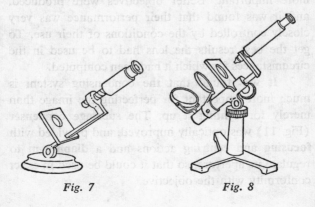

Fig. 7   *Fig. 8*

had to be strongly illuminated; a **mirror** to direct the light, and a **condensing lens** to concentrate it were added (Fig. 8).

5. A great advance in optical theory occurred in 1832, enabling lenses to be combined in such a way that sharper images free from coloured fringes could be produced. These were the **achromatic** and **aplanatic** objectives of J. J. Lister (Fig. 9). Higher magnification became worth while.

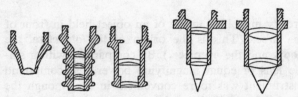

Fig. 9. Evolution of the ach-   Fig. 10. Achromatic objec-
    romatic objective.   tives of different aperture.

6. It was found that magnification is not the chief factor in revealing fine detail in a specimen. Light-grasp, or **aperture** (Fig. 10), in the objective is much more important. Better objectives were produced, and it was found that their performance was very closely controlled by the conditions of their use. To get the best results the lens had to be used in the circumstances for which it had been computed.

7. It was found that the condensing system is much more important for perfecting the image than merely for lighting it up. The **substage condenser** (Fig. 11) was optically improved, and provided with **focusing** and **centring** actions and a **diaphragm** to regulate its aperture, so that it could be used in exact conformity with the objective.

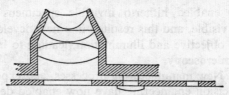

Fig. 11. Early substage condenser [Kingsley type] c. 1840.

8. The German scientist Ernst Abbe investigated the optics of the microscope scientifically, and developed a theoretical basis for objective design. This led to a spectacular improvement in performance, and microscope design became a science instead of an art.

9. Mechanical improvements were introduced as manufacturing processes developed. The microscope became a standardized instrument, with no prospect of further development (Fig. 12).

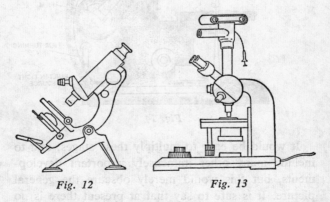

Fig. 12   *Fig. 13*

10. The Dutch scientist Frederick Zernike investigated the action of the specimen on light, and produced a new optical technique—**phase contrast—**

which enabled hitherto invisible specimens to be
made visible, and this resulted in a new development
of the objective and illumination, leading to **interference microscopy.**

11. New manufacturing processes, and experience
of precision engineering have now enabled designers
to break away from the conventional microscope,
and design afresh (Figs. 13 and 14) without the limitations previously controlling what could actually be
produced in practice.

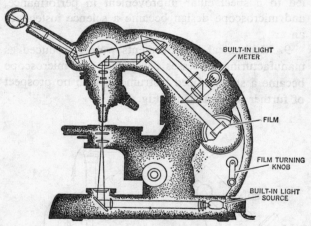

Fig. 14

It would be easy to multiply these stages, and to
include many other immensely important developments, but this would merely obscure the general
picture. It is safe to say that at present there is no
tendency to regard the microscope as a finally-
evolved piece of apparatus, and further changes are
to be expected.

Description of Typical Stand

For introductory purposes it is best to consider the simplest form of microscope in ordinary use. With the exception of mounted magnifying glasses, to which the name "Simple Microscope" is properly applied, the simplest conventional microscope is that designed for the examination of transparent specimens, which are commonly mounted on glass **slides** $3'' \times 1''$ to support them, and protected above by a very thin glass **coverslip.**

The slide is supported on the microscope (Fig. 15) by a **stage,** which is pierced by an aperture to transmit the light, and viewed through a tube rigidly aligned with its axis perpendicular to the surface of the stage. This tube, or **body,** carries two separate systems of lenses which, by their co-operative action, reveal the structure of the object—the **objective** which screws into the lower end of the tube, and the eyepiece or **ocular,** which slides into the upper end.

In many microscopes, the tube itself, though incapable of lateral movement, can be moved along its axis to bring the lenses into adjustment, sliding in a groove in a supporting **arm.** The arm, to which the stage is attached, extends beyond it to carry a **mirror** which has a flat and a concave surface for use alternatively, and is capable of being turned in any direction to reflect light up through the specimen. In many newer microscopes (Fig. 15), the arm is immovable. The lenses are brought into adjustment by axial movement of the stage assembly.

On all but the very simplest microscopes the sliding bearing carrying the tube is continued down the extension of the arm to accommodate a further system

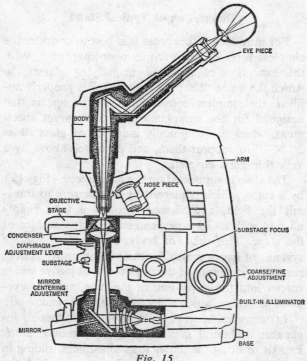

EYE PIECE

BODY

ARM

NOSE PIECE

OBJECTIVE
STAGE

CONDENSER

DIAPHRAGM
ADJUSTMENT LEVER

SUBSTAGE

SUBSTAGE FOCUS

COARSE/FINE
ADJUSTMENT

MIRROR
CENTERING
ADJUSTMENT

BUILT-IN ILLUMINATOR

MIRROR

BASE

Fig. 15

of lenses—the **substage condenser**—which is located
between the mirror and the stage and used to illuminate the object.

The three systems of lenses are all mounted with
their optic axes in the same straight line, and all act
independently in the production of the image.

The Objective

The **objective** is the most important of these lens
systems, and the mechanical adjustments of the mi-

croscope are all provided to permit its efficient action. Upon it depends the amount of detail which can be displayed, and it also contributes the greater part of the necessary magnification. It has already been explained that these two factors—the resolving power, or ability to discriminate between adjacent details, and the magnification—commonly go hand in hand as a matter of convenience, but are actually distinct, and must not be confused. The objective receives light from the specimen on the stage, and projects an image of it (Fig. 16) into a position near the top of

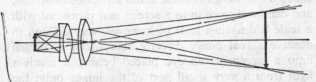

Fig. 16. Projection of a magnified image of the object by microscope objective.

the body tube. It thus acts like the lens of a photographic enlarger or cinema projector, forming an inverted image which is larger than the object because it is proportionately further from the lens.

In the case of a cinema projector, this image is thrown onto a white screen and viewed directly, the texture of the surface of the screen diffusing the light which it receives over a wide field of view, from any part of which it is possible to see the complete picture. It is clear from this that the picture has to be very brilliant to provide sufficient light at all positions in front of it, but from the point of view of a single observer, all the diffused light which does not enter the pupils of his eyes is entirely wasted; the picture he sees is in no way improved by its availability to the rest of the audience. For a solitary observer this is, therefore, an extravagant method of display, as well as technically inconvenient, and it is only used in microscopy in the circumstances in which it is used in entertainment—the exhibition to several people simultaneously of a simple specimen.

The photographic enlarger is in a rather different category where illumination is concerned, as the light acts on the sensitive film, and any reflected light is wasted; ideally, the film should absorb all it receives and reflect nothing, whilst the cinema screen should diffusely reflect everything and absorb nothing.

An image of this type, projected through a lens, is called a **real image** because it has an actual existence, and can be thrown onto a screen and measured with a scale laid against it. It is, however, very difficult to observe a real image directly unless it is projected onto a screen, as an eye placed behind it receives light from a very small part of the image only; the rest is spread out over the observer's face.

The Eyepiece

By the use of a second lens in the fashion of a magnifying glass to observe the image projected by the first, it is possible to view the entire image; the second lens collects the diverging light rays which have already formed the primary image, and causes them to converge again so that they may all enter the eye (Fig. 17).

This results in a second magnification, which leads to the name **Compound Microscope.** The total magnification is the product of the primary, or objective, magnification, and the secondary, or eyepiece, magnification. It is important to notice that the eyepiece does not contribute anything new to the image, but simply amplifies it, spreading out the details already

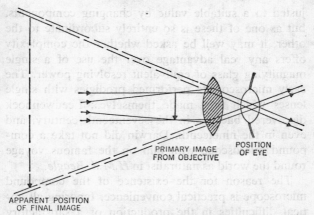

PRIMARY IMAGE
FROM OBJECTIVE

POSITION
OF EYE

APPARENT POSITION
OF FINAL IMAGE

Fig. 17. The eyepiece directs light from the primary image into the eye, and also amplifies it.

present sufficiently for the eye to distinguish them; the details are provided by the specimen, the objective, and the illumination.

If amplification is carried to excess, by the use of too powerful an eyepiece, the final image becomes vague and woolly; the effect is comparable with that of magnifying a photograph until the grain becomes evident and confuses the picture. There is no advantage in magnifying something in which all the detail present is already clearly visible; more minute examination requires the inclusion of still finer detail in the original image, and this can only be supplied by changing the objective for a more powerful one.

Reason for Compound Construction

The division of the total magnifying power into separate stages permits it to be conveniently adjusted to a suitable value by changing components, but as one of these is so entirely subordinate to the other, it may well be asked whether the complexity offers any real advantage over the use of a single magnifying glass of equivalent resolving power. The early microscopists performed prodigies with single lenses which they made themselves; Leeuwenhoek discovered bacteria in the seventeenth century, and even in the nineteenth, Darwin did not take a compound microscope with him on the famous voyage round the world as naturalist in *H.M.S. Beagle.*

The reason for the existence of the compound microscope is practical convenience. Ignoring all optical difficulties in the production of a satisfactory image, the magnifying power of a single lens depends on the curvature of its faces. Once a lens of given diameter has attained the form of a sphere, it has reached its limit in this respect, and greater convexity can only be obtained with a smaller lens. As the magnification is raised, the lens thus becomes smaller and smaller, until it is narrower than the pupil of the eye. Not only does this reduce the amount of light capable of entering the eye, but also, even when the lens is held close against the eyelashes, with the object equally close to the lens on the far side, the field of view is minute, and imperfections in the lens of the eye interfere with the view. Any manipulation of the specimen becomes practically impossible, so that it is necessary simply to accept whatever happens to come into view. An additional feature of great inconvenience is that the image seen is **virtual**; it has no real existence, and is as inaccessible as the reflection in a mirror. There is no possibility of bringing it into focus on a graticule and thus making measurements with such an image, nor making it visible to a second observer by projecting it on a screen.

In fact, the simple microscope begins to become awkward at a magnification of 20x, and where dissection is necessary, the modern practice is to use a compound stereoscopic binocular microscope with image-erecting prisms; this provides magnifications up to 120x with ample working distance and eye-space. The compound microscope provides a greater range of useful magnification, a longer working distance, and room for the observer to wear spectacles if necessary; in addition, it can be used as a binocular with both eyes. It is possible also to apply micrometers, scales, and other devices to the image without difficulty, and to project or photograph it.

Substage Condenser

The third system of lenses, the **substage condenser,** controls the quality of the image formed by the objective; it provides the raw material which the objective works up after the specimen has acted upon it. This component governs not only the visibility of the specimen, but also the resolution to a considerable extent, and it is only a slight exaggeration to say that with it the microscopist manufactures the image he sees. Its action is duplex; it provides a cone of light to exploit the full power of the objective, and it enhances the intensity of illumination of the specimen to provide a picture bright enough when enlarged.

It follows from this that the use of the substage condenser, and the selection of a suitable pattern for the work in hand, is an important consideration. Much can be gained or lost by the manner in which it is adjusted, so that it is well worth paying particular attention to this component, which is more completely under the control of the user than the objective or eyepiece, which once in focus work as well as the condenser permits.

Stand Controls

The mechanical portions of the microscope, collectively termed the **stand,** consist of the stage, tube, arm and mirror already mentioned, the base which supports the whole, and the mechanical adjustments. The base of older microscopes is commonly a horseshoe-shaped casting, which rests upon the bench top at three widely-spaced points to ensure stability. An ascending portion of the casting is usually attached to the arm through a stiff hinge joint, so that the microscope can be conveniently inclined, though sometimes arm and base may be integrally constructed to provide greater rigidity, and the optic axis of the microscope inclined by means of prisms. A great deal of emotion has been expended on the shape of the microscope base; formerly British makers used a tripod support, and European makers a horseshoe with a pillar. It was contended that the tripod was more rigid, and provided greater steadiness when the instrument was inclined. Actually, provided the hinge joint is placed well between the supporting pads of the base and sufficiently high on the arm to reach the centre of gravity of the part inclined, there is nothing to choose between the two types. In fact, the true tripod has long disappeared.[1]

The mechanical adjustments of the microscope are

[1] Most modern makers now offer inclined bodies and circular or rectangular bases.

normally four in number. Three of these are in constant use in connection with the focusing of the lenses, and involve longitudinal movements of the body and condenser mount, and the fourth ensures the correct alignment of the substage condenser with the optic axis.

The body tube of older microscopes or the stage assembly of some newer designs is moved up and down its slide by a rack and pinion, controlled by large knurled heads to provide a delicate control. The teeth of the rack are diagonal, and those of the pinion spirally cut; this ensures that a number of teeth are in engagement simultaneously, so that the movement is smooth and exact.

This is the **coarse adjustment,** and has a range of several inches to provide scope for focusing a variety of objectives. A good coarse adjustment is astoundingly responsive, but for focusing high power lenses where a movement of one thousandth of a millimetre (one micron, μ) is significant, more precise means are essential.

Accordingly a **fine adjustment** is provided as well, producing a very slow action through some form of lever, screw, or gearing. The controlling heads of this movement are usually smaller than those of the coarse adjustment, and are provided with a scale and index measuring displacements of the body by units of the order of 0.001 mm.

Formerly the fine adjustment tended to be the weakest point of most microscope designs, but it has benefited by developments in engineering practice, and those produced now are satisfactory. It should nevertheless be treated with the care befitting a delicate mechanism, as its action makes or mars the stand; an erratic or worn fine adjustment is unusable.

The substage condenser is focused by a rack and pinion action exactly like the coarse adjustment, though with a restricted range; sometimes students' stands have a worm-and-nut action instead, but this is not suitable for serious work with the condenser.

Substage Centration

The microscopes now being offered by leading makers can be provided with a centring action to enable the substage to be aligned with the objective in use. This centring adjustment is not a confession of slipshod work, but a recognition that the user may need to employ components which have not been aligned during the process of manufacture with those supplied originally. It is absolutely essential for high-power dark-field illumination or phase-contrast work with any microscope.

The ordinary centring system consists of a pair of radially-directed screws set in the rim of the substage mount at right-angles to each other. These bear on an inner sleeve which carries the condenser itself, and force it against a spring on the opposite side of the mounting (Fig. 88A). By adjusting these two screws it is thus possible to move the axis of the inner sleeve laterally, and to align the condenser with the objective.

The use of a **revolving nosepiece** to carry two, three, or four objectives simultaneously on the body of the microscope, ready for alternative use, is universal. It is a safe rule always to rotate a revolving nosepiece clockwise; some of the older ones are thrown out of alignment if turned backwards. The objectives are attached in order of increasing power, so that an object may be scanned with a lower power and examined with a higher one without delay. The objectives are adjusted to ensure that in these circumstances they are approximately aligned, and display the same field of view at approximately the correct focus, thus reducing the time taken by readjusting when a change is made. Objectives not belonging to the set for which the nosepiece was originally adjusted naturally do not automatically agree in centration or focus with the regular set.

Tube Length

The distance between the bottom of the nosepiece and the upper end of the microscope tube is called the **mechanical tube length,** (Fig. 18) and is conventionally fixed at 160 mm by all makers except Leitz, who uses 170 mm. Objectives are adjusted to give their best performance when separated by this distance from the eyepiece flange and focused on a

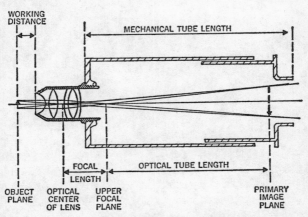

Fig. 18. The relationships of the working distance, focal length, and tube length of an objective.

specimen covered by a standard coverslip. As coverslips vary considerably, it is sometimes necessary with high-powered objectives to compensate for abnormalities by using a tube length other than the standard, and monocular microscopes generally have the body made in telescopic form to permit this. Binocular microscope bodies lack this **draw-tube,** which cannot readily be incorporated in the design.

Chapter Three

THE MICROSCOPE—
MANIPULATION AND CARE

Assembly

A microscope is normally supplied in a fitted case, with two or more objectives, two eyepieces, and a condenser. The first thing to be done with a new instrument is to ensure that it can be removed from the case and replaced with certainty and without force. Although this may seem childish, it sometimes proves unexpectedly difficult, as the case is made to support the stand and protect it from shocks, and consequently has various buffers fixed to bear against the stand from the sides and back, and it is common to find another on the door. If the instrument is not initially arranged in a vertical position with the body racked into the right position, it may prove impossible to insert it between the buffers or to shut the door. In former days there was a tendency to make the cabinet fit its contents so very exactly that packing away assumed the nature of a Chinese puzzle; every component had to be removed from the stand and trapped separately in its individual bed, and the tube racked down to its stop, with neither objective nor eyepiece in place. Nowadays a more practical, if less intellectually amusing, attitude prevails, and the instrument can be put away fully assembled for immediate use. This is an enormous advantage; not only can it be used without delay, but the wear and risk of attaching objectives, and the entry of dust into the empty tube, are eliminated.

A new microscope is sometimes located in its box by a screw inserted up through the bottom of the case into the base of the instrument, as a safeguard during transport. This screw, after removal, may be retained against future need, and the hole in the box plugged with a small cork to exclude dust.

The instrument as removed commonly has the mirror and substage condenser in position, whilst the objectives and eyepieces are fitted separately in a drawer. The stand should be examined in detail to attain familiarity with its arrangement; the controls are generally as shown in Fig. 15, but some new models are quite differently built. With a new microscope the instruction manual should be consulted, as it is possible that some of the adjustments are locked or tightened up for security during transport.

The following instructions are intended to avoid damage during the period whilst familiarity is being attained, and to promote good habits. Familiarity is entirely desirable, but it must on no account be permitted to degenerate into contempt; there is no place in microscope manipulation for force, carelessness, or ostentatious capability.

Before the instrument is assembled for use, it is well to blow sharply down the tube, to dislodge any dust which may be present, and then to insert an eyepiece to ensure that no more enters; an eyepiece should always be left in position for this reason.

The objectives should be attached to the nosepiece; each is removed from its box in turn, and screwed into place. Formerly objectives were held in brass boxes by an R.M.S. thread in the lid, but this practice has long been discarded, and the objective is now trapped by a projecting flange between the cap and the body of the box as these are screwed together. It is thus free to fall out of the box as this is undone, and therefore it is a sound practice to hold the box down on the table whilst unscrewing the cap. It is prudent to use both hands when screwing the lens into its position in the nosepiece, one locating it by means of the flange in the appropriate hole, and the other turning the barrel of the lens to engage the thread; this reduces the danger of dropping it.

Objective Nomenclature

The power of the objectives will be indicated on them. The modern usage is to give a magnification figure, or else the focal length of the component, which is preferable as being less ambiguous. At one time makers used individual series of numbers or letters of arbitrary value, and some makers reinforce the engraved data with a code of coloured rings to make recognition easier. If there is any doubt about the identities of the objectives, it may be taken that the larger front lens belongs to the lower power. Objectives commonly supplied with microscopes have the following characteristics:—

Characteristics of Objectives commonly encountered

Category	Magnification	Focal Length mm	ins.	N.A.	Working Distance mm
Lowest power	3.5x	24	1	0.08	24
Very low power	5x	32	1½	0.1	22
Low power	10x	16	⅔	0.25	8
Medium power	40x	4	⅛	0.65	0.5
High power	90x	2	1/12	1.25	0.2
					(in oil)

In many cases the apertures in the nosepiece are marked to correspond with the appropriate objectives, to ensure that alignment shall be retained when the nosepiece is rotated. In any case, when three objectives are mounted on the nosepiece, they should be arranged for future convenience in ascending order of power as the nosepiece is rotated clockwise. Once they are screwed into position, firmly but not excessively tightly, they should be left there and not returned to their boxes after use; nosepieces are dustproof, and removal of the objectives only results in wear, exposure to dust, and risk of accident.

If there is a high power objective in the set, it should not be mounted in the first instance, as this objective requires experience before it can be used without risk of damage. It is likely that where two only are supplied, these will be the 10x and 40x, as these are practically ubiquitous, and the eyepieces commonly used with them are the 6x and 10x, giving total magnifications of approximately 60x, 100x, 240x, and 400x.

Setting Up and Focusing

1. For the initial essay, it will be found easiest to use daylight to illuminate the specimen, though there are sufficient disadvantages about daylight to drive a microscopist to the use of a lamp for routine work.

2. The microscope should be set upon a rigid table close to, and facing, a window, but out of direct sunlight, and should be inclined at a comfortable angle.

3. The substage condenser, where present, should be racked up until its top is practically flush with

the surface of the stage, and the iris diaphragm below it opened half-way.

4. A slide should now be placed on the stage with the specimen upwards, and immediately above the condenser. If a highly coloured prepared slide is available, it is to be preferred for this purpose, but, failing this, a sprinkling of any opaque powder—pepper, cocoa, etc.—in a drop of water covered by a thin coverslip, will serve. The sole requirement in the case of this particular specimen is that it should be incapable of being overlooked or missed.

5. The mirror of the microscope should be turned about, flat side uppermost (if the microscope has a substage condenser), until the specimen, viewed from above, is seen to be lit up.

6. The body, with the low power objective (10x, 16 mm) in line with the tube, should now be racked down until the front of the objective is about ⅛″ from the slide. Looking into the eyepiece, the microscope body is racked slowly *up;* the field of view, coloured by the dye of the object, should differentiate and the specimen come into view. If it fails to do so, the probable reason is that the objective was not close enough to the slide at first. It is sometimes helpful if the slide is moved about a little on the stage, as the eye is very sensitive to movements, which can be detected before the object is properly in focus; it is thus possible to ensure that there is something actually in the field of view to focus upon.

7. Once the object is in sharp focus, it is possible to adjust the illumination so that it can be properly seen. This is best effected by focusing the condenser to throw an image of the source of light into the specimen, which has the effect of making the object behave to some extent as though it were self-lumi-

nous. As in the present case the source of light is the open sky, the simplest way of focusing the condenser at its best position is to move the mirror until some object such as a window-bar comes into the field of view, and to rack the condenser down until this is sharply in focus with the specimen, after which the mirror can be re-adjusted to exclude telescopic objects from the field of view. See Chapters 7 and 8 for the correct adjustment of artificial light sources.

8. The lenses are now properly focused, but two steps in the process of adjustment remain. First, the centration of the condenser must be checked, and secondly the amount of light must be regulated. For both purposes the eyepiece is removed, and the back lens of the objective observed, the head being held well back from the open tube, so that the eye is looking axially down it. The iris diaphragm is slowly closed and opened; it will be seen imaged in the back of the objective as a bright disc which expands and contracts as the condenser aperture is changed. The bright disc should lie centrally in the objective, indicating that the condenser and objective have their optic axes in line. If this is the case, attention can be directed towards regulating the amount of light. If, however, the illuminated disc is not centrally placed in the objective—that is, if it does not reach the margin of the objective lens everywhere around it at the same opening—adjustment is required, and the radial screws of the substage centring action are used to bring it to the proper position. Microscopes lacking this adjustment are usually carefully centred beforehand; with an old instrument it is worth trying the effect of rotating the condenser in its sleeve, as this usually provides a certain amount of adjustment in this case.

9. Having checked the centration of the condenser, the iris should be closed until the illuminated area covers about $\frac{9}{10}$ of the objective lens, and the eyepiece re-inserted. The effect of alterations in the iris setting can now be studied. If the aperture is increased, the image becomes brighter, but tends to lose contrast owing to glare. If it is closed, the image becomes duller and less incisive, but colourless features such as fibres may become apparent which were invisible when the diaphragm was further open.

It is essential that it should be understood that the use of the diaphragm is not to regulate the brightness of the image; it controls the angularity of the illumination—the width of the base of the cone of light impinging on the specimen and expanding thence into the objective—and thus the visibility, resolution, and nature of the image. By and large, it should be kept as far open as glare will allow within the circle of the back lens of the objective, though it may be somewhat closed to reveal the presence of colourless structures. These should not be examined with it shut, as their outlines are then expanded to a misleading extent.

Low Power Survey

With the low power objective in position and the illumination properly arranged, the specimen should be carefully examined; this is an exercise in manipulation and appreciation, and the exact nature of the object is not immediately important. Sir E. Ray Lankester always made his students study air bubbles as their very first specimen, to ensure that they would never fail to identify them subsequently. Air bubbles are easily obtained by merely licking a slide and placing a coverslip in position; their peculiarity lies in the differing appearances which they exhibit as the focal level is changed from a point below to a point above them, and which are due to lenticular effects of the spherical bubble. This is an excellent object for training purposes, and should be carefully studied by every beginner.

It will be found that movement of the slide on the stage is initially awkward because under the microscope the movements are reversed, and although with familiarity this is no longer noticed, it makes manipulation somewhat erratic at first. It is valuable to be able to manipulate the slide by direct finger movements, and this should be practised without the assistance of a mechanical stage (if one is fitted). The clips usually supplied to keep the slide firmly in place on the stage are generally a nuisance; if the stage is horizontal they may be turned out of position or unplugged from their sockets. Their up-turned ends can in some cases touch the front of the objective lens, which is of course to be avoided. Systematic examination of a slide on a stage fitted only with clips is not easy, but can be facilitated by using a strip of card or other material a little thicker than a slide, trapped horizontally under the clips to provide a guiding ledge along which the slide is slid (Fig. 19).

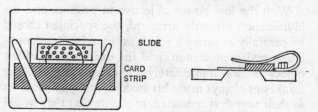

Fig. 19. Use of a strip of card under the stage clips to facilitate specimen manipulation.

The guide is itself moved up or down at the end of each traverse. The formal arrangement of this device, called a **sliding bar,** is now most uncommon but still very useful.

Generally speaking, a microscopist uses his low power for the initial appreciation of the general arrangement of a specimen, and his medium or high power for critical examination. This low power approach should never be neglected even in routine examinations of uniform specimens like blood smears; it provides the opportunity for a quick look round, which may reveal a tell-tale detail or condition of the specimen unlikely to be noticed in the more restricted condition of high power examination, when the object cannot be seen as a whole, or as an extended area.

It is important to bear in mind that if the necessary detail in the object can be made out under the low power, there is no advantage, and a great many disadvantages, in using a higher one. The low power objective is more convenient to use, embracing a wider field with a greater depth of focus, and giving an image more readily appreciated than the medium or the high power lenses, which are essentially adapted to the examination of local details. Low power objectives can be used with powerful eyepieces more effectively than high powers can, as their resolving power is disproportionately higher than their magnification, thus providing a wider margin for mere amplification. It is a safe rule to keep the total magnification within a limit of $1000 \times$ N.A., so that a 10x objective of 0.25 N.A. can be used with eyepieces up to 25x. However, the limits of permissible magnification are somewhat academic, and are based on the behaviour of very good eyes; those who are limited to a somewhat lower intrinsic resolving ability may find that a higher magnification is desirable. In actual fact the components in normal use do not allow their users to exceed or even attain the $1000 \times$ N.A. limit except when the microscope is used to project an image.

Setting Up Medium Power

The use of the medium power objective is not equivalent merely to changing to a higher power eyepiece. It provides an image containing much more detail than the low power image, more highly magnified, and consequently much more restricted both in area and in thickness. This makes it somewhat harder to focus, and as the increased resolving power automatically involves a much reduced working distance, which will be less than one millimetre, there is a distinct danger that the lens may be racked through the slide if great care is not taken.

All manufacturers adopt the practice of adjusting their own objectives to be approximately in focus when the nosepiece is rotated to bring the 40x into focus after the 10x, thus saving both anxiety and time. Nevertheless, objectives by different makers are not automatically par-focal, and the practice is not of very long standing, so that it is reckless to assume that an unfamiliar microscope will conform to it, or that it can safely be relied upon except when examining ordinary thin specimens on slides with thin coverslips.

When it is first carried out, therefore, *the change from low to medium power should be observed from the side of the microscope,* to ensure that clearance really exists to allow the medium power lens to be rotated into place.

Before making the change, notice the working distance of the low power objective—that is, the distance from its front to the coverslip—as a guide for the future. This working distance must never be confused with the focal length, which is not measured from the front of the lens and is generally much greater.

1. It is instructive to remove the eyepiece and to open the iris diaphragm until it just coincides with the margin of the objective lens; this is properly carried out with the eye close to the end of the tube, to obtain the proper perspective.

2a. Now replace the eyepiece, and rotate the nosepiece, clockwise, watching from the side, till the medium power lens clicks into position. Normally there is no difficulty about this, but if the lens cannot be brought into place, check that the slide is the correct way up, and if so, rack the body up just far enough to allow the lens to come home. In all probability it will swing smoothly into place, and the image be brought sharply into focus by a slight movement of the fine adjustment. The direction of movement of this adjustment and its approximate extent should be noted as a guide for future use. If, on the other hand, no image is visible at all, try moving the slide a minute amount; it is possible that the field of view coincides with a gap in the specimen.

2b. If it has been necessary to disturb the original focus to align the lens, start with the objective as close to the slide as possible, and rack slowly up, looking down the eyepiece, until the image appears, using the fine adjustment to perfect the focus. The position of the objective with respect to the slide should be noticed, for future reference.

The precautions suggested are advisable for an absolute beginner lacking any help at his first attempt, and will rapidly be relegated to limbo as experience is gained. At the same time it must be emphasized that carelessness, familiar contempt, or light-heartedness in bringing lenses into focus is utterly misplaced. It is admitted that most microscopes are par-focal, and some even made fool-proof by recoil springs to protect the objectives, but there are still many instruments which are not so convenient, and it is a golden rule to assume that any unfamiliar microscope is likely to be cranky.

3. When the specimen has been brought into focus as described, the eyepiece should be removed again and the back of the objective examined. It will be found that the diaphragm setting which just filled the low power objective with light illuminates only about a third of the back lens of the higher power. The difference in this respect represents the increase in resolving power now available. The diaphragm lever should be slowly moved to enlarge the disc of light, which ought to expand to fill the objective.

4a. If this disc appears eccentric in the medium power objective, it indicates that this objective does not have the same optical axis as the low power, with which the condenser was aligned, and this requires compensation. The standard procedure is to centre the condenser by adjusting its radial screws until the objective is centrally illuminated, but before doing so it is worth checking that the objectives are in their intended seatings in the nosepiece, and that the latter has been turned fully into its proper position. Unless the nosepiece is turned into correct register, which will be indicated by feeling its spring catch engage and click into place, the axis of the objective will be grossly displaced sideways.

4b. Centration with the higher powers is more important than with the low, and the simple examination down the open tube is hardly exact enough. It is possible to use a cap on the tube, pierced by a small central hole, to ensure that the view down the tube is axial, but the more usual method is to keep the eyepiece in place and to look through the microscope whilst lowering the condenser *with its iris closed as completely as possible*.

If this is done, an image of the iris eventually comes into focus. It will appear as a polygonal bright area on a dark background, usually showing blue and red fringes on opposite sides; these indicate that the light is travelling obliquely through the microscope.

4c. The substage centring screws are now adjusted until the polygon is centrally placed in the field of view; it may be helpful to move the slide aside to obtain an unobstructed view of the iris image whilst this is carried out. The coloured fringes will be practically eliminated when the iris is centred. When a lamp is used instead of the sky as a light-source, it may be necessary to adjust the mirror to remove the colour fringes, but in the present case there is no question of the condenser being unevenly lit; the sky is wide enough to eliminate this.

Ideally, of course, it is the objective which should be moved, to keep the line condenser-objective-eyepiece straight when a change of objective bends it, and this can be effected in some microscopes. However, the substage centring device can be made much more massive than an objective centring device, and much more reliable. In either case the actual effect of the process in making the eyepiece extra-axial is imperceptible, as the angular displacement of different objectives, measured from the eyepiece, is minute, and the lateral displacement of the objective negligible with respect to the diameter of the eyepiece. One professional method of centring the different objectives of a set to a particular nosepiece depends on filing away metal at one side of the seatings until the objectives when screwed home show fields of view as identical as possible; this obviously involves what is in effect moving the eyepiece across the field of view.

5. Once the condenser has been centred, the condenser may be racked up again to its position of focus, using the window-bar as a reference point as before. See Chapters 7 and 8 for the adjustment necessary for use with artificial illumination. It is worth focusing it first with the diaphragm practically closed, and then opening the iris and checking the focus. When the condenser is of the simple type called an Abbe Illuminator, it will be found that the mere opening of the iris diaphragm brings the condenser focal level nearer to the top of the condenser than when the iris is closed. The reason for this is that the rays passing marginally through the Abbe Illuminator come to a focus lower than those passing through it axially.

The better types of condensers are corrected to eliminate this effect, which wastes a great deal of light and is liable to cause glare, and they maintain an identical focus with any objective in use. Nevertheless, the Abbe Illuminator is used on more than 99 per cent of all microscopes, as it is cheap, and does not require careful adjustment. Generally speaking it is adequate for objectives up to 0.85

N.A. (the routine 40x and lower magnifications) but is incapable of fully meeting the requirements of more powerful objectives, which are therefore, used inefficiently with this condenser.

Medium Power Characteristics

When the objective and condenser have been focused on the specimen, it should be examined to demonstrate the delicacy of movement required to bring any desired feature to the centre of the field of view, and also the greatly reduced focal depth available with the medium power objective in comparison with the low power. With a specimen of any appreciable thickness, the fine adjustment will be in constant use, and the field of view will be found to contain a whole series of superimposed structures which come into view and vanish as the focus is changed from the top of the specimen to the bottom.

As the result of geometrical optics, the axial magnification of a specimen is the square of its magnification transversely, so that a cube magnified twice along a side appears as a pillar four times as high as wide (Fig. 20). This means that the structure of the specimen in depth must be inferred; it cannot

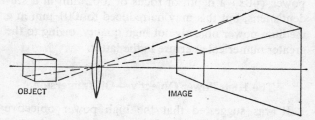

OBJECT   IMAGE

Fig. 20. Distortion in the image of a thick specimen.

be comprehended at a glance if the thickness exceeds the focal depth of the objective. The eye has a certain latitude of focus, or **accommodation,** and can appreciate to some extent the dimensional relationships of the structure seen, whilst the ability to focus the objective at different levels extends this facility mechanically, so that an appreciation of the object can be obtained by assembling various views. A photograph of the field of view lacks both of these facilities; what is out of focus is invisible. For the majority of purposes, therefore, a drawing of the microscopical appearance is more instructive than a photograph, and the microscopist should always make a drawing in preference; it is not old-fashioned to do so, and the microscopist who can attain fluency with his pencil is the better able to record his observations and instruct others.

The depth of focus of an objective—the range over which a point is, loosely speaking, visible as a point and not a disc, depends inversely on the numerical aperture of the objective, and thus inversely also on its resolving power. The higher the N.A.—the more obliquely the objective views the object—the more abruptly it will be defined in depth and the more exactly in plan (Fig. 21). The low power (10x) may have a depth of focus of 0.01 mm, and the medium power (40x) a depth of focus of .002 mm in a student's lens, but this may be reduced to .001 mm in a medium power objective of high quality, owing to the greater numerical aperture in the latter.

The High Power Objective—Oil Immersion

It was suggested that the high power objective should not be mounted in the first instance, in case

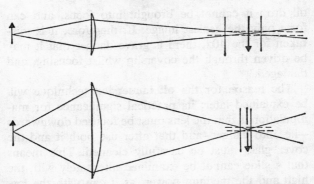

Fig. 21. The diminution of focal depth by increased N.A.

of accident, but it is convenient to indicate the particular points about its use which justify this advice.

First, it must be understood that this lens is designed for the examination of very thin specimens, such as dried films of blood in which bacteria, parasites or abnormalities of the individual corpuscles are to be studied. Its normal objects are the individual components of a single cell, and its depth of focus, about .0004 mm, enables it to cut optical sections of a single white blood corpuscle, and to distinguish between bacteria above, in, and below it. The lens can be used with thin tissue sections, but is rarely necessary in this case unless the cell contents rather than cell aggregates are under consideration.

It is the standard working lens in bacteriology and cytology, but is seldom used in biological studies except for the traditional sport of diatom examination. It is not only delicate and expensive, as all objectives are, but also has an extremely short working distance, and *is used with its front lens dipping in a drop of special oil applied to the coverslip.* Without this oil drop it cannot be brought into focus, and can form only the haziest image. If, therefore, it is mistaken for the 40x, there is grave danger that it may be driven through the coverslip whilst focusing, and damaged.[1]

The reason for the **oil immersion** technique will be explained later; its practical significance for manipulation is that the lens must be focused downwards —i.e. into danger—and that after use both it and the cover glass must be carefully cleaned. This means that a slide cannot be examined alternately with the high and the medium power, as it can with the low and medium, because the oil has to be wiped off, and except in the case of permanently mounted slides this destroys the specimen. In ordinary biological work, where most objects are examined in water under a cover, there is rarely any advantage in using the oil immersion objective, as the problems which arise are those of visibility, in which it offers no advantage, and not resolution, in which it is supreme. Unless particular circumstances demand it, it should be reserved for permanently-mounted slides or for stained and dried smears.

Before putting the immersion objective into use, the nature of the immersion oil and the method of removing it after use must be understood. It has become evident that a warning must be given that the term "oil immersion" refers to the use of oil between the lens and the specimen, and not in any circumstances to the use of oil *inside* the mount of the objective; this unprecedented error has been reported several times recently, but its source has not so far been revealed.

[1] Many makers engrave a black line around the lower end of the oil immersion objective to distinguish it.

Immersion oil was formerly a cedar wood oil thickened by the addition of Canada Balsam to produce a solution having the optical properties of glass, the refractive index and the dispersion being matched to those of the cover glass and the front lens of the objective. It had the disadvantage of drying hard in air if left long exposed, and is now largely supplanted by synthetic liquids free from this objection. These non-drying oils have their own disadvantages, however, as their optical properties differ somewhat from the original, and various makers design their objectives to work with particular oils, so that the combination of the correct oil and objective is important for the best results. There is a risk that the use of an inappropriate liquid may weaken the cement around the front lens of an objective, and oil may actually enter the mount around it if this occurs. This is probably a remote possibility where immersion oils themselves are concerned, but is real enough where cleaning liquids are considered.

The objective must be carefully cleaned after use; even if the oil remains fluid, even if it does not dissolve or penetrate the mounting, even if it does not oxidize and attack the glass eventually, it will collect dust, and its sticky surface will be a nuisance. The oil is wiped off with lens paper, and a piece of lens paper moistened—not soaked—with xylene is used to remove its traces, after which a dry piece of lens paper will complete the process. Many years ago the use of alcohol was recommended by an author in a moment of aberration; this unfortunate mistake ought at all costs to be avoided as, although modern mounts may be capable of resisting it, it is liable to soften the cement used for securing the front lenses of older objectives. Any mount old or new should be safe with the normal xylene.

The presence of dried oil or Canada Balsam on the front lenses of the medium or high power objectives is a perennial nuisance in laboratories. It may be removed in the manner described; some people prefer to immerse the front of the lens in cedar wood oil to soften the dried matter, but on balance it is safer to rub it with a moist lens paper from the start—it is not the lens surface which is being rubbed, but the pad of resin covering it.

Setting Up High Power

1. To bring the high power objective into use, following the 40x medium power, first check that the microscope tube, if it is adjustable in length, is set to 160 mm overall (or 170 mm if the objective is by Leitz), rack the body up a centimetre or so, check that the fine adjustment is well away from its bottom limit, and rotate the high power into position.

2. A small bead of immersion oil is now applied to the top of the cover glass where the condenser is illuminating it. The drop should be about 3 mm across, convex and compact. As a rule the oil is kept in a bottle having a rod extending from the bottom of the stopper into the contents, and this is used to transfer a suitable quantity. The use of a common plastic oil-can has been recommended for handling immersion oil, and should be excellent.

3. The objective is now lowered into the oil by using the coarse adjustment *whilst the operation is watched from the side*. From the point at which contact is made with the oil, the fine adjustment is used, and the eye applied to the eyepiece. Remember that

the objective is being focused downwards, and that there is very little margin between the focal position and actual contact with the cover glass. If the specimen is well-defined, there should be no real difficulty about locating the focus, but with isolated particles it is easy to meet a bare field of view.

For this reason it is best to select a well-covered field of view before abandoning the 40x objective, and to use a low power eyepiece at first with the oil immersion, in order to secure the widest possible field of view. It is a good plan to arrange that the sub-stage condenser is projecting into the plane of the specimen a well-defined image of the light source, as the gradual differentiation of the bright field into a luminous structure can be an invaluable guide in difficult cases. The use of a disc of wire gauze in front of the lamp, focused on the specimen during the transition to form an unmistakable target and then removed, is useful at first.

4. Once the specimen has been located, the illumination must be adjusted, to ensure that the condenser is centred to the objective. Obviously it is impossible to do this by racking up the body to locate the image of the closed iris diaphragm, and if the condenser is lowered to obtain the same effect, it will usually be found that the iris cannot be closed enough to come into view. If a lamp fitted with a diaphragm is being used, it will have been centred to the 40x objective, and any difference in centration between this and the high power can be compensated by using the condenser centring screws to bring the closed lamp condenser into the centre of the field of view.

5. The adjustment of the substage iris is not as critical in the case of the oil immersion objective as

with the medium power, because unless an **oil immersion condenser** is used, it will be impossible to fill the back lens of the objective with light. Opening the diaphragm will cause the illuminated disc seen in the objective to expand to a certain extent, but beyond this it has no effect; the immersion objective can accept a wider cone of light than the dry condenser can give.

This completes the elementary instruction for bringing the microscope into use for the first time. The detailed consideration at a further stage of the objective, condenser, and eyepiece will provide criteria to enable the user to judge whether he is using his apparatus to the best advantage, but it is necessary as an introductory measure to deal with the basic processes of image formation.

As previously mentioned, proper cleaning of the lenses is extremely important. It is always good practice to have handy some device for removing dust before actually wiping the lenses with lens paper. Otherwise, it is possible that grit from the air will scratch the lenses as you try to clean them. Many devices for this purpose are available from local camera stores; a small, good quality camel's hair brush will be satisfactory and almost any small blowing bulb, such as a baby enema syringe, available in most drugstores, will do the job nicely.

For cleaning the lenses, one should always use good quality lens paper or an old, clean linen handkerchief; other tissues or cloth may contain harmful abrasive materials. It is best to fold the lens paper several times to avoid smudging the lenses with oils and moisture from the fingers. To remove oils or grease smudges, it is advisable to dampen a lens tissue slightly with xylene or toluene. Excess solvent is to be avoided since it may soften or dissolve lens mounting cements.

Sometimes the most difficult problem in cleaning is to find out where the dirt is. If the dirt moves as you rotate the ocular, or if it moves with the specimen slide or comes into and out of focus with adjustment of the condenser, then you have solved your problem. Grit eventually gets into the moving surfaces of microscopes which are continually exposed in dusty rooms and, when the microscope does not focus smoothly or when it becomes difficult to focus, the microscope should be disassembled and cleaned. The process should be repeated frequently in areas of great potential harm to microscopes, such as a geology laboratory. However, while it is not recommended for inexperienced persons to disassemble and lubricate microscopes, every microscope owner should recognize that this process is necessary.

Once the microscope is disassembled but before lubrication, the old grease should be removed with a soft cloth and xylene. The various microscope manufacturers have their own recommended lubricants but a useful one can be prepared by making a 1 : 1 mixture of melted pure petroleum jelly and pure lanolin.

Chapter Four

LIGHT AND THE OBJECTIVE—I

Wave Theory

It was pointed out above that a student's medium power 40x has a greater depth of focus than one of the finest type, and that this depends on the N.A. of the lens. Attention has been drawn to the independent existence of the phenomena of visibility and resolution, and it has been emphasized that the image of a point is inevitably a disc, smaller or larger according to circumstances, but a disc nevertheless.

The connection between these apparently diverse items is the behaviour of light, which is naturally the controlling factor in instruments which are dependent on it. Optics is a wide subject, but the microscopist need only be concerned with some characteristics of it in relationship to its behaviour in microscopical circumstances, and the treatment may therefore be kept as simple as will serve to allow him to work efficiently. An intelligent appreciation of its behaviour will enable him to decide what practices are advantageous, and what are best avoided or are acceptable with reservations.

It is common knowledge in these days that light may be considered as a wave motion, similar to, but on a smaller scale than, the waves of radio transmission. These waves are not mechanical vibrations passing along the straight rays, but oscillations in the electro-magnetic conditions existing there, which can, however, be treated for purposes of calculation as though they were material undulations.

We know that if we gradually raise the temperature of an object, its particles are forced into a state of increasingly rapid vibration, emitting energy in a form we appreciate initially as heat, and then, as the object becomes incandescent, as light, first dull red, and then, as the temperature rises, orange, yellow, and ultimately a blinding white. This corresponds to the emission of a collection of vibrations of various frequencies, to one group of which the eye happens to be sensitive, and the presence of a conventional selection of these is recognized as white light, though there is nothing actually coloured in the vibrations; the various colours into which white light can be divided represent a kind of code by which the eye and brain identify them.

The frequency of vibration of a particular light wave is its characteristic feature, which cannot be changed, and which remains constant until the wave ceases to exist (Fig. 22A). Separate waves having identical frequencies can combine to form a single wave representing the mathematical sum of the components (Fig. 22B—E), but if their frequencies are not absolutely identical they behave as though in isolation.

If light rays can combine and reinforce or suppress each other—in other words, **interfere** with each other —they are said to be **coherent,** and such rays originate from the same luminous source; rays from different sources, or different particles in the same source, cannot interfere in this manner, and are said to be **non-coherent.**

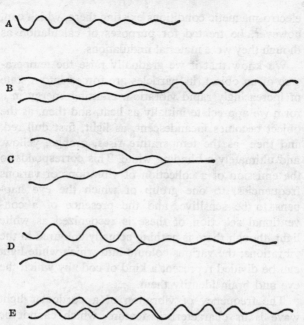

Fig. 22. Behaviour of coherent light waves. A) The frequency of a light ray remains constant, though the amplitude may diminish as the ray is absorbed. B) Combination of identical rays in similar phase relationship doubles amplitude. C) Destructive interference of identical rays in reverse phase: amplitudes cancel out. D) Interference of rays of same frequency but different amplitudes, in reverse phase, causes change of amplitude but not phase. E) Interference of identical rays, one quarter out of phase, changes phase but not amplitude.

Refraction

In air, light travels at about 186,000 miles per second, but in liquid or solid media such as water, glass, or oil, it travels more slowly; the ratio of the speed in air to that in the medium is termed the

refractive index of the substance, and thus, taking the speed in a vacuum as unity, the refractive index of a substance will be greater, commonly lying in the range of 1.33 to 1.7, though values in excess of 2.0 exist.

The frequency of vibration of light is unaffected by this slowing down of transmission, so that there are more waves per unit distance in more highly refractile media than in lesser—in other words, the wavelength is diminished. A ray passing from air into glass, water, and air again, may be imagined as being contracted in this fashion (Fig. 23).

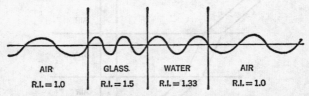

| AIR | GLASS | WATER | AIR |
| R.I. = 1.0 | R.I. = 1.5 | R.I. = 1.33 | R.I. = 1.0 |

Fig. 23. Variation of wavelength with light velocity.

This property of refraction enables the ordinarily straight course of light rays to be changed by the use of components such as lenses and prisms, which slow down one side of a beam of light before the other, and so deflect its mean path of travel (Fig. 24).

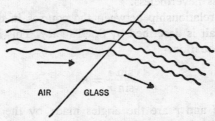

AIR   GLASS

Fig. 24. Bending of a light beam by refraction.

A beam of light falling obliquely on the upper surface of a thick glass block will have its velocity reduced first on the underside, and will consequently incline more steeply downwards as it enters the glass. The converse occurs at the lower surface of the glass, where the light emerges. If the two faces of the block are parallel, the light will resume its original direction, though the actual path of the emergent ray will be off-set with respect to the entrant ray (Fig. 25). If the light beam were **normally incident**—i.e.

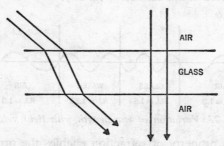

AIR

GLASS

AIR

Fig. 25. A light ray is laterally deflected by a sheet of glass unless it passes perpendicularly through it.

perpendicular to the surface of the glass—its course would be unchanged, though it would be retarded in the glass nevertheless.

The relationship between the courses in the glass and in air is described by **Snell's Law.** This is stated as an equation,

$$\frac{\sin i}{\sin r} = n,$$

where i and r are the angles made by the incident and refracted rays with the perpendicular at the point of entry, and *n* is the refractive index of the glass.

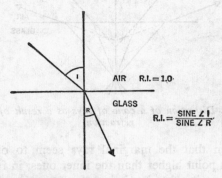

Fig. 26. Snell's Law. Refraction of a light ray at a glass surface.

This relationship can be easily deduced from that between the velocities in the two media.

Spherical Aberration—Its Correction

For microscopical purposes it is easiest to consider the rays of light as originating in a glass block, and emerging into air. A luminous point in the glass block will emit light rays in all directions, and those which are emitted upwards will constitute a conical beam of light which will fan out as they emerge into air, each ray diverging from the optic axis to the extent provided for by the formula, the outer ones most and the inner ones least (Fig. 27).

It follows from this that the cone of rays which has emerged from the glass consists of rays which could not have achieved their new courses if they had come from a source in the air; tracing them back, it will

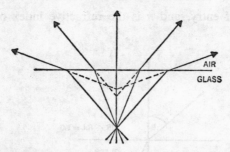

Fig. 27. Distortion of a cone of rays as a result of surface refraction.

be seen that the marginal rays seem to originate from a point higher than the inner ones; in fact, the spread as it exists in air is not a true cone, but the familiar caustic curve associated with burning glasses.

If these rays then enter a lens, they will not be brought to a true focus unless the lens is made to counteract this condition, which is known as **spherical aberration** (Fig. 28B), and to bring them all into directions consistent with their having originated at a point (Fig. 28C).

This **spherical correction** can only be effected by taking into account the thickness of glass from which the rays have emerged; if the thickness is altered, the extent of the spherical aberration is altered also, and consequently requires a different degree of correction (Fig. 28D).

It might be thought that by using an objective with a flat front, so that it was parallel with the upper surface of the glass, the spherical aberration could be corrected completely and inevitably, as the rays resume their original courses once they enter the glass of the lens. Consideration will show, however, that this is not the case; individual rays will certainly

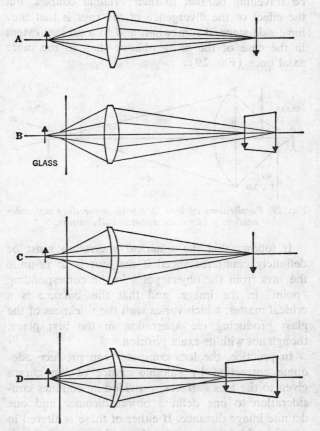

Fig. 28. Spherical aberration and its compensation. A) Ideal relationship with no spherical aberration. B) Introduction of spherical aberration in image by spherical aberration at object. C) Spherical aberration in image eliminated by compensating for original error at object. D) Change in original error at object re-introduces error in image.

be travelling parallel to their original courses, but the effect of the divergence of the rays is that they have side-stepped somewhat, and to a greater extent in the case of the more oblique than in the more axial ones (Fig. 29).

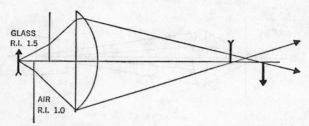

Fig. 29. Parallelisms of lens face with cover does not automatically eliminate spherical aberration.

It follows that the spherical aberration must be definitely counterbalanced if the lens is to re-unite the rays from the object point in the corresponding "point" in the image, and that this balance is a critical matter, which varies with the thickness of the glass producing the aberration in the first place, though not with its exact position.

In practice, the lens computer can produce adequate correction by suitable choice of the curves given to the lens surfaces, provided he confines consideration to one definite object distance and one definite image distance. If either of these is altered in practice, the other must be changed to a corresponding extent to recover the balance, and bring the rays to a common focus.

Theory of Homogeneous Immersion

It will be noticed that this aberration arises from the refraction of the rays leaving the glass and enter-

ing air before returning to glass in the lens. By eliminating the air gap, this initial spherical aberration
can also be eliminated.

This is done by using the principle of **homogeneous
immersion,** in which the space between the front of
the lens and the top of the cover glass is filled by a
fluid, usually a special oil, having the same refractive index as the glass, as has already been described.
In this case (Fig. 30) the first refraction which the

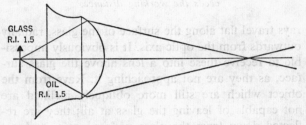

*Fig. 30. Elimination of spherical aberration at the object by
homogeneous immersion.*

rays undergo is at the upper surface of the lens, and
this is arranged so that it occurs **aplanatically**—that
is, without introducing spherical aberration. In these
circumstances the thickness of the glass covering the
object is immaterial; the lens is maintained at its
proper distance, and if the gap contains more glass,
it contains less oil, and vice versa. Provided that the
cover glass is not so thick that the lens cannot be
brought to the focal position, it does not matter at
all (Fig. 31).

A further advantage is gained by homogeneous
immersion. Rays diverging from an object in glass,
and being refracted progressively more obliquely as
they impinge on the free surface and pass into the
air, ultimately attain the condition in which the outer

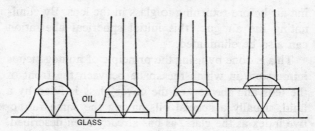

Fig. 31. In homogeneous immersion the lens-object distance is constant, and cover thickness is immaterial until it exceeds the working distance.

rays travel flat along the surface of the glass, radially outwards from the optic axis. It is obviously impossible to receive these into a lens above the glass surface, as they are not approaching it. Rays from the object which are still more obliquely inclined are not capable of leaving the glass at all; they are reflected down from the glass surface again, and not transmitted across the boundary (Fig. 32A).

The limiting condition, in which the obliquity of the rays in glass is such that after refraction they make an angle with the axis in air of 90° is called the **critical angle.** The value of the critical angle depends on the refractive index of the glass; as the sine of 90° is 1.0, the corresponding angle in glass is therefore given (by Snell's Law) by the relationship $1 = n \sin r$. Consequently any ray for which $\sin r$ exceeds $1/n$ cannot be transmitted through the boundary, and is lost to the image.

The homogeneous immersion principle rescues these rays, which are transmitted through the oil to the lens, and moreover transmitted without deviation from their original courses (Fig. 32B). This is a point of great importance, as it enables the lens to

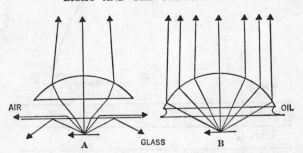

Fig. 32. Greater availability of rays from an object in homogeneous immersion owing to elimination of losses due to refractive effects.

utilize rays which would in air embrace a total angle of more than 180°. In fact, the "effective obliquity" of a ray is determined by the refractive index of the medium in which it is travelling, so that optically, rays making axial angles of 72° in air, 40.5° in water, and 38.75° in oil or glass are exactly equivalent.

It must be noticed that if a set of rays has to traverse a series of different refractive indices, for instance when an object is mounted in water between a glass slide and cover, and examined with a dry objective, the maximum angle of transmission is limited by the lowest refractive index in its path (Fig. 33). In the series given, the glass slide could transmit more oblique rays than the water, which would therefore reflect these, passing on to the coverslip rays up to its own limit, which would in turn be reduced by the air gap. There is thus no advantage in using a homogeneous immersion objective on a specimen mounted in air under a cover (Fig. 34), and a watery mountant will to some extent limit the potential light-grasp of such an objective. In both cases the automatic correction of spherical aberration introduced

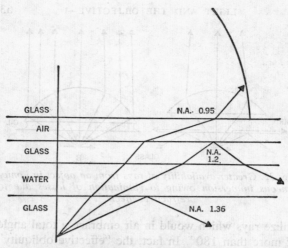

Fig. 33. Limiting effect of minimum refractive index on the transmission of oblique rays through layers of different materials into an objective.

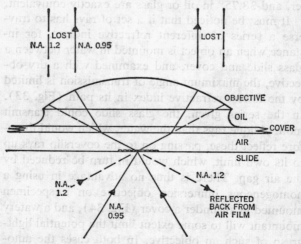

Fig. 34. Inutility of using an oil-immersion objective with a dry-mounted specimen.

by the coverslip will not take place, and other means
will have to be adopted to deal with it.

The importance of the ability to collect as much
light from the specimen as possible lies in its effect
on the resolution of which the objective is capable.
So far the numerical aperture has been mentioned
without being exactly defined, but it must now be
considered in detail.

Numerical Aperture of an Objective

The **numerical aperture** of a microscope objective
is defined as the value $n \sin u$, where n is the refractive index in the object space—i.e. on the receiving
side of the objective—and u is the axial angle made
by the most oblique ray passing from the object to
the image (Fig. 35B).

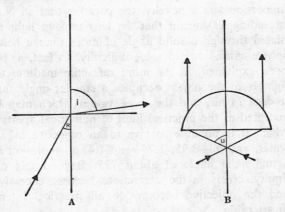

*Fig. 35. Correspondence between critical angle in refraction
and limiting value of N.A. in an objective.*

It thus corresponds exactly to the critical condition in plane refraction (Fig. 35A); the angle u,

which is controlled by the mechanical design of the lens, corresponds to the critical value of the angle r in the earlier examples, and we have seen that the refractive index determines the actual physical magnitude of this angle, and therefore has to be specified.

Just as a given medium cannot transmit rays for which the value $n \sin r$ reaches a value of n, because this corresponds to an axial angle of 90° and hence no forward movement of the rays, so an objective is limited to a value of n for its numerical aperture, since the rays must travel towards it from the afferent medium. This imposes theoretical limits for the numerical aperture of dry, water and oil immersion objectives of 1.0, 1.33, and 1.52. It may be noticed that as N.A. 1.0 corresponds to a full right-angle, a dry objective of N.A. 1.0 would receive all the light radiated forward by an object. When homogeneous immersion was a novelty, the paradox that an N.A. exceeding 1.0 meant that the lens took in light radiated through a solid angle of more than a hemisphere caused considerable difficulty. In fact, as has been explained, in the more refractile medium the light from the object occupies a smaller angle than it does in air, and the three values which may be accepted as the practical limit of numerical aperture which it is possible to give to an objective in air, water, and oil—0.95, 1.25, and 1.4, all represent an actual axial angle of about 72°. Beyond this obliquity, errors in the corrections become excessive, and the objective becomes to all practical intents unworkable.

It is important that it should be realized that the ability of a lens to work at a high numerical aperture is dependant on its receiving rays up to its limiting obliquity. If these are stopped, for example by a diaphragm either in front of or behind the lens, the working aperture is reduced to a corresponding extent. Immersion objectives require a liquid or solid contact with the object if they are to work at full aperture; the interposition of an air layer automatically excludes all rays corresponding to N.A. 1.0 and over (Fig. 36).

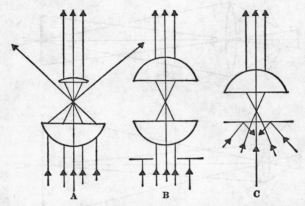

Fig. 36. Limitation of working N.A. by A) Restriction of objective diameter. B) Restriction of illuminating cone. C) Lack of immersion contact with specimen.

The concept of Numerical Aperture has now been explained in some detail, and it remains to show what it signifies in practice, why it is worth using immersion lenses to obtain a higher value, and why, if a high value is beneficial, manufacturers offer lenses with different numerical apertures at all.

The importance of N.A. in the behaviour of an objective is supreme. The resolving power depends directly on the N.A. in linear relationship; using light of the same colour, N.A. 1.3 will resolve detail twice as fine as N.A. 0.65, and the lens of 0.3

N.A. will resolve twice the detail of the lens of 0.15 N.A.

The N.A. of a lens depends on its axial angle, so that a lens of longer focal length may have as large an N.A. as a narrower one of shorter focal length (Fig. 37), which will project a larger image. It is,

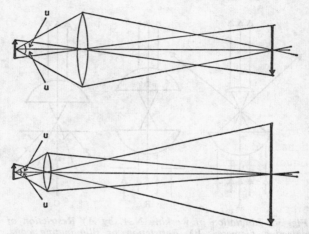

Fig. 37. Objectives of different focal lengths but equal N.A.

therefore, independent of magnification, and the word "power", referring to lenses, is better applied to the N.A., which governs the resolving power, than to the magnification, which is a means to an end of no intrinsic importance.

From what has been said of the need to use a magnification capable of enlarging all the detail in an image to visible proportions, it will be obvious that the ultimate value usefully attainable depends only on the numerical aperture of the objective.

A value of 1000 × N.A. is commonly quoted as

the useful limit, beyond which further magnification is "empty", as nothing further is brought to view. It is possible, however, to argue that greater magnification ensures that nothing is overlooked. As a rule, the available magnification is limited by the apparatus available, and it is unusual for this to provide excessive magnifications in visual use, though they can be obtained easily in photography by increasing the projection distance or by subsequent enlargement of the negative.

Whilst the use of a high N.A. provides greater resolution, it has practical drawbacks, so that the unconsidered use of the highest N.A. available is misguided. It will be appreciated that the effective use of very oblique rays involves increasingly stringent corrections for spherical aberration, as slight variations in cover thickness produce much greater effects. Equally, the more oblique the rays, the more exactly is the focal plane defined, so that objects above and below it are not seen. The conditions essential for producing a relatively high N.A. produce also considerable curvature of the image. In short, the higher the numerical aperture of an objective, the more inconvenient it becomes; spherical correction becomes critical, working distance diminishes, the field of view becomes excessively curved, and the price rises sharply. For critical work in the hands of an expert, the higher the N.A. the better, but for general use there are advantages in a lens with a lower N.A., less sensitive to its working conditions, with a flattish field of view of wider extent, and a longer working distance.

The size of the standardized objective thread provides the ultimate limit to the diameter of the lens which can be accommodated in an objective, and the

obliquity of the rays needed to fill this lens depends
on its focal length, and hence magnification. The
numerical apertures of objectives commonly supplied at the present time are as follows:—

Category	Magnification	Focal Length mm	ins.	N.A.	Working Distance mm
Lowest power	3.5x	24	1	0.08	24
Very low power	5x	32	1½	0.1	22
Low power	10x	16	⅔	0.25	8
Medium power (dry)	40x	4	⅙	0.65	1.0
Medium power (dry)	"	"	"	0.85	0.5
Medium power (oil immersion)	50x	3.25	⅐	1.0	0.5
High power	90x	2	¹⁄₁₂	1.3	0.2

It will be seen that there are three objectives in the
medium power category, two 40x's and the 50x oil
immersion. The first 40x lens is predominantly designed to have a long working distance, a flat field
of view, and to be insensitive to variations in cover
glass thickness. It is easy to use, and suitable for
routine work. The 40x lens of .85 N.A. is capable of
much greater resolution, and consequently of being
used with a more powerful eyepiece without deterioration of the image. Using the 1000 × N.A. rule,
the lens of 0.65 N.A could attain a total magnification of 650x, and the other of 850x, which correspond to the use of eyepieces magnifying 16x and
21x respectively. As in ordinary circumstances eyepieces with a power exceeding 10x are unusual, this
may not appear to show a convincing superiority
for the 0.85 objective, but it must be remembered
that in photography these magnification values could

easily be exceeded with a 6x ocular and a long camera.

With the 0.85 N.A. objective, it is necessary to pay attention to the thickness of the cover glass. Most makers correct their objectives for a cover thickness of 0.17[1] mm when the body of the microscope is 160 mm long from the objective shoulder to the eye-piece flange, and departures from the assumed cover thickness can be compensated by adjusting this tube length, reducing it for a thick cover, and increasing it for a thin one (see p. 83). This adjustment is made by inspection of the image, and not by measurement, and requires a certain amount of experience. However, the lens will provide a reasonably good image if worked at a set tube length of 160 mm and not adjusted.

The third medium power lens listed is the immersion of 1.0 N.A. This has a better resolving power than the last-mentioned dry lens, a long working distance, and is insensitive to cover thickness. In fact, it is an extremely competent objective in all respects. The aperture is deliberately kept low for an oil immersion, because the objective is primarily intended for use with a dark-field illuminator (see p. 181), but oil immersion objectives of 60x have been constructed with apertures up to 1.4, for use instead of the conventional 90x with the same aperture. In this case, the outstanding advantage is the increase in working distance, which is 0.05 mm with the 90x of this aperture, but 0.13 mm with the 60x.

Where the aperture has to be kept down to 1.0 to permit dark-field illumination, and an immersion objective, with its insensitivity to cover thickness, is desirable, there is much to be said in favour of the

[1] No. 1½ coverslips.

relatively unknown **water immersion** objective, which is available with a magnification of 50x.

Water immersion objectives were popular before the homogeneous immersion principle was enunciated, and are still manufactured, though seldom advertised. A good modern water immersion lens has an aperture of 1.0 N.A. and is much less sensitive to cover glass thickness than a dry lens of comparable aperture. It has the great advantage that the distilled water used as an immersion fluid may be blotted off the cover without disturbing the specimen if it is necessary to revert to a lower power objective again. The water immersion objective is not to be compared with the oil immersion objective where resolution is important, but where the case is one of visibility, in routine work, there is a lot to be said in its favour. This is particularly the case in biology, where specimens in water are studied under a cover glass. There is no optical advantage in the oil immersion in these circumstances, as although the water of the mount will transmit rays oblique enough to fill the aperture of the objective, the presence of the water layer introduces the spherical aberration which the oil is intended to eliminate.

The high power objective, the 90x oil immersion, is now given an N.A. of about 1.25 for students and 1.3 for routine use. The relationship of resolution to N.A. is strictly *pro rata,* and the delicacy and awkwardness of objectives with apertures of 1.4 (and in the case of one Powell & Lealand type, 1.5) is not considered commensurate with their increased resolution nowadays, when other means are available to elucidate difficult specimens. Visual microscopy is now settled down in a comfortable middle

age, with optics chosen for their all-round convenience, and does not indulge in the inconvenient vir-
tuosity of its youth; today's problems are those of
visibility, not resolution, and whilst we may hold the
apochromatic 1.4 N.A. in respect, better resolution
can be attained photographically with less exacting
lenses.

It will be noticed from an examination of the values in the list of objectives that the lower power objectives have a higher aperture for their magnification
than the higher power ones. This implies that greater
eyepiece magnification can be used with the low powers than with the higher powers; the limits on the
1000 × N.A. formula are as follows:—

Objective magnification	N.A.	Total magnification (1000 × N.A.)	Maximum ocular magnification
5x	0.1	100	20x
10x	.25	250	25x
40x	.65	650	16x
	.85	850	21x
90x	1.3	1300	14.5x

In practice this theoretical limitation is less exact
than it might appear, as so much depends on the
quality of the objective and of the eyepiece; individual lenses vary quite markedly, some being capable
of unexpectedly fine performance, and others beginning to fail before the anticipated limit is reached.
In any case, if the objective is to be used to its
limit, care must be taken over its adjustment to the
particular specimen in view; coverslips are not uniformly thick, and the spherical correction cannot be
made on a rule of thumb basis.

The Exit Pupil

Where visual microscopy is concerned, there is a minimum as well as a maximum permissible value of the magnification. The light proceeding up the microscope body is brought to a focus in the eyepiece, which produces a virtual image of the object at a distance Dv from the eye which is nominally taken as 250 mm. The light from this image, or apparently from this image, passes through the eyepiece and converges into a small disc immediately above the eye lens before diverging again (Fig. 38). This

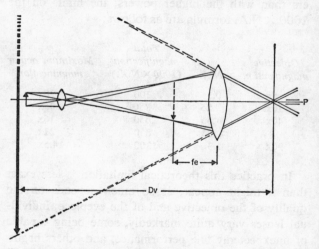

Fig. 38. The exit pupil is an image of the objective aperture projected by the eyepiece.

small disc is termed the **exit pupil** of the microscope; it is actually an image of the objective aperture, and its diameter thus depends on the N.A. of the objective and the total magnification.

The actual relationship between the diameter DR of the exit pupil (or **Ramsden circle**) and the magnification and N.A. is given by the formula

$$DR = \frac{2\ Dv.\ \text{N.A.,}}{M} \quad \text{or,} \quad DR = \frac{500 \times \text{N.A.}}{M}$$

The significance of this is that the diameter of the pupil of the eye in bright light is approximately 3 mm, and unless it can receive the whole of the exit pupil, the effective N.A. will be cut down. In this case, the peripheral parts of the image will be formed by asymmetrical lighting, and movement of the eye will cause movement of the image.

To reduce the diameter of the exit pupil to reasonable size, a magnification of at least 170 times the N.A. is required (Fig. 39).

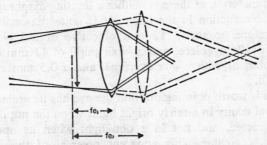

Fig. 39. Influence of eyepiece magnification on size of exit pupil and height of eyepoint.

It may be reduced considerably before the conventional limit of $1000 \times$ N.A. is reached, but if the exit pupil is made too small, physical defects in the eye itself begin to become apparent as shadows on the image. To avoid this, and to allow for slight movements of the eye during observation, an exit pupil diameter of 1.5 mm will be found to be a good working compromise. This corresponds to the following eyepieces for the various objectives:—

Objective	N.A.	Useful minimum magnification	Eyepiece
5x	0.1	33	7x
10x	.25	84	8x
40x	.65	215	5x
	.85	280	7x
90x	1.3	430	5x

Eyepieces of lower magnification than 4x are extremely rare, so that it follows that this condition is not likely to be violated, as even the 4x ocular would not produce an exit pupil of excessive size.

This minimum magnification applies only in visual work; in photomicrography the size of the exit pupil is immaterial, as there is neither a limiting diaphragm nor obstruction in the beam. It is quite reasonable, therefore, to use a 12 mm objective of .65 N.A. with a 2x eyepiece, and an exit pupil of 11 mm diameter, or with a 40x eyepiece, and a 0.5 mm exit pupil.

It is worth pointing out that the eye has its greatest visual acuity in a fairly bright light, when the pupil is contracted, and not in a dim light when its aperture is greater. This apparent reversal of the expected behaviour arises from limitations on the correction of the lens of the eye; its aberrations increase with its aperture.

Chapter Five

LIGHT AND THE OBJECTIVE—II

The Ideal Image. Phase Maintenance

We have so far considered the effects of refraction, and the aberrations which are introduced, without reference to the wave structure of the light, except for the influence of refractive index on wavelength. The wave structure of light is, however, profoundly important, and must be borne in mind by both the user of the microscope and the designer.

A small source of light—an incandescent particle—will radiate light in all directions, and the vibrations will be identical in nature in all directions of propagation; they will be equal in frequency, and the vibrations will all be in time with each other. In other words, at all points on a spherical surface with the radiating point as centre, the waves will be in the same condition as far as displacement is concerned, though not necessarily vibrating in the same plane. The waves are entirely coherent, and can be combined mathematically.

Theoretically, if all the radiating waves could be collected in an identical manner, so that they were converging upon a point instead of leaving one, the original particle could be reproduced in complete identity with the original.

This state of affairs can never be achieved, for the reason that wherever the original particle may be re-imaged, some of the original radiation will be travelling in the opposite direction, and will be lost. Only a part of the radiated light can thus be used for producing an image, and the image will fail exactly to reproduce the original because part of the light is missing; in other words, some of the original information has been lost.

The more that can be collected by the optical system and transmitted to the image—that is, the greater the N.A. of the optical system—the more exactly will the object be delineated, assuming that the spherical correction of the system is perfect. This supposes that the rays all come to a focus at a point and not in the general neighbourhood of a point, and that they arrive in the condition of uniform phase relationship in which they started.

The importance of the phase relationship depends on the ability of coherent rays to interact; mathematically, the result of combining two coherent wave-trains is additive, so that the amplitude of the resultant waves depends on their mutual phase difference. Two equal waves exactly in phase will equal one of double the amplitude of either, and two equal waves completely out of phase will neutralize each other (Fig. 22). Waves partly out of phase combine to form a resultant of intermediate phase, but the same wavelength, and with an amplitude which depends on their sum, and waves of unequal amplitude produce a resultant with an amplitude greater than or smaller than that of either component, according to the phase relationship.

Waves of different frequencies act in entire independence; they do not combine in any way, and preserve their identities.

Translated from the mathematical to the practical

plane, the amplitude of the waves represents the brightness of the light, and the relationship is based on a square law; twice the amplitude represents four times the brightness. Thus two coherent waves of equal amplitude can quadruple the intensity of the light, but two non-coherent waves can only double it, as these act independently.

Restricting consideration to the case of coherent waves, which are uniting to form an image of one particle, it is, therefore, important to ensure that the combination should take place with the same phase conditions as existed originally. If this is not the case, the image will differ from the object to a greater or lesser extent; at the worst, the image will be extirpated by the interference of the waves.

The achievement of the condition of phase integrity, which may be called the **conventional condition**, is obviously secured if all the rays from object to image have performed the same number of vibrations on the way—in other words, if their optical paths are equal. Axial rays through a lens travel the shortest distance through air and the longest through glass, whilst marginal rays travel further in air and a shorter distance in glass (Fig. 40).

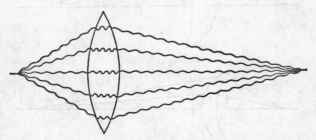

Fig. 40. Ideally, all light-paths from object to image are equal.

It may be considered that the problem of bringing together converging rays to the pitch of accuracy represented by one quarter of a wavelength—the accepted criterion—represents the triumphant achievement of the incredible, but it is possible to examine objectives on an **interferometer,** in which the pattern of the phases is made visible, and to have ocular proof not only that the mathematical basis has practical meaning, but also that the specified result can be attained. The manufacturing processes involved are the most exacting of any to be carried out on a regular production basis.

Coma

The provision of equal optical paths for the marginal and axial rays must be effected in such a way that the ratio of object-lens distance to lens-image distance is constant for all rays, or different parts of the lens will project superimposed images of different sizes (Fig. 41). This causes a loss of defini-

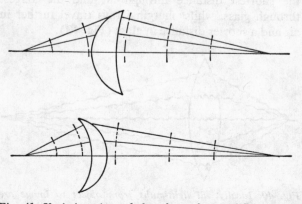

Fig. 41. Variations in path length producing differences in magnification of image projected by different parts of a lens.

tion except at a point on the axis of the lens; an object point lying off the axis will give rise to an image which will be a radial line—the condition known as **coma.**

In terms of lens design, success in meeting the requirement is satisfied by fulfilling a relationship called the **sine condition,** which is the business of the lens computer, who combines this with the correction of spherical aberration in such a manner that both are simultaneously satisfied as far as possible.

It will be obvious that this entails computations of great complexity, so that it is reasonable to accept that no general solution to the problem is possible, but only one for a specific set of conditions. If a given object distance and a given value for the spherical aberration are accepted, an optimum result can be calculated for a specific distance to the primary image, but if any of these is not in fact correct in practice, the quality of the image will suffer (Fig. 28).

Cover Glass Errors

With objectives of long working distance and low aperture the effect is not serious, but when the thickness of the coverslip is appreciable with respect to the working distance, discrepancies between the assumed and true values of its thickness and refractive index have a significant effect.

High power objectives have, therefore, to be compensated for such variations, and this can be effected by providing them with a **correction collar** (Fig. 42), which varies the separation between the front lenses, which provide the magnification, and the posterior lenses which correct the errors in the front ones. This enables the user to adjust the lens to the exact conditions in which it is being used (Fig. 43), though a

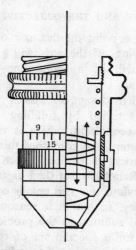

Fig. 42. A 4 mm objective of 0.95 N.A. with correction collar marked in terms of coverslip thickness.

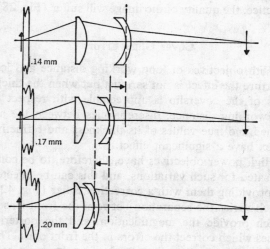

Fig. 43. Action of a correction collar (diagrammatic).

similar result can be attained by adjusting the tube length—an alternative recommended by some manufacturers.

Correction by Tube Length and by Correction Collar

The adjustment of an objective becomes necessary at numerical apertures of 0.85 and over with dry lenses, and in the prevailing conditions when the optical constants of cover glass and of immersion oil are somewhat chaotic it is well worth carrying out even with oil-immersions, as true homogeneous immersion is now unusual.

The principle involved depends on the fact that the objective is designed for particular conditions of object and image distance (Fig. 44A). The two are interlocked by the ability to vary the distance at which the primary image is formed by varying the distance of the objective from the object; if the length of the body is 190 mm instead of 160 (Fig. 44D), the object-objective distance must be decreased to secure the extra projection distance, and if the tube is used in the closed position at 140 mm instead, the object-objective distance will be correspondingly greater (Fig. 44E). Owing to the obliquity of the rays reaching the objective from the object, the differences in the distance, though minute, have a profound effect on the spherical correction, whilst the relatively slight obliquity of the rays passing up to the eyepiece render this side of the system insensitive to such influences, though the magnification will be changed as the tube length is altered.

The principle is exactly the same with a correction collar, though the influence on magnification is then much less.

similar result can be attained by adjusting the tube-length—an alternative recommended by some manufacturers.

Correction by Tube-length and by Collection Collar

The adjustment of an objective is-times necessary at numerical apertures of 0.65 and above by similar lenses . . . the eye-ining condenses which the optical consists of cover glass on of imperfection of its conservator chromatic it is well worth carrying out even with illuminations to order not correct in the margin . . . to the image that.

The principle involved depends on the fact that the objective is designed for particular positions of object and image (distance 'L'). . . . Thus, if it is desired to possible the corrections of an image while the primary image is formed by varying the distance of the objective from the object together with of the body . . . mm instead of the (Fig. 44A). . . . so that the correct . . . . secure the same objective distance and if the tube is used . the closed position at 160 mm then at the object-objective distance will be considerably greater . . . . rays reaching the objective from the object, the differences in the distance, though minute, have a profound effect on the spherical correction and the relatively the majority of the rays passing into the eyepiece under this side of the system can involve to such influences . . . the magnification may be changed . . . .

the principle is exactly the same with a correction collar, though the influence on magnification is then much less.

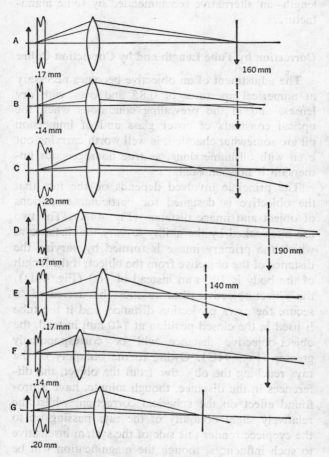

Fig. 44. Correction by tube-length adjustment.

It is thus possible to introduce a condition of spherical aberration in the image by working the objective at an incorrect tube length, or with an incorrect collar setting. If the mounting of the specimen differs from that which was foreseen by the computer, this will, as was explained earlier, introduce spherical aberration to an extent different from his allowance, and the image will suffer (Fig. 44B, C). By using the deliberate spherical error due to a deviation from the official tube length to counteract that due to a casual deviation from the anticipated mounting conditions, the situation can be restored to normal (Fig. 44F, G). With a thick cover glass, producing extra aberration, the requisite degree of excessive correction is produced by shortening the tube length, and vice versa.

The difficulty in practice lies in recognizing the extent of the adjustment required. This is not a circumstance in which measurement of coverslips with a micrometer, which is sometimes advised, is of any appreciable use, as an unknown thickness of mounting material will be involved also. There is a school of thought which prefers to use a tube of invariable length as an insurance against wrongly executed correction, or the ill-advised use of the draw-tube for changing the magnification. Some microscopists insist that no variation in the image is discernable, and that objective correction is a fad, but these are demonstrably wrong except when the objective is either of relatively low aperture or else so badly constructed that there is no scope for obtaining a good image from it.

In fact, it will be found that with a sensitive objective the proper correction can make the difference between visibility and invisibility in a delicate specimen, but only experience will help the microscopist to carry out the adjustment, and the subjective element is apt to be considerable. The operation is best practised by using an artificial object which provides minute points of light, as it is much easier to appreciate the appearances which are shown by a luminous point than by an opaque one against a luminous field, and the complete opacity of the opaque point is important.

Luminous points are provided by using a slide coated with silver by chemical means, and having covers of different thicknesses cemented to the silver, which will inevitably be pierced by microscopic perforations; such slides are available commercially, or may be prepared. An alternative is to use a film of the dense stain **nigrosin,** which will also present minute flaws in its expanse, and another is to use light reflected from minute beads of mercury deposited from vapour or from a drop of mercury struck by a spring to disperse it, though this last is inconvenient, and hard to cover without spoiling. It is also rather risky to expose the microscope to possible contact with mercury.

A suitable object may be extemporized by using a high power dark-field illuminator to illuminate an ordinary slide, slightly dusty on its upper surface, with a coverslip laid on it.

In any case, the slide is examined to locate a minute bright point, too small to have any discernible structure, which may be considered as an artificial star—a source of light without size.

This point is brought into focus, and will be seen to form an image consisting of a central disc of light surrounded by rings of light which decrease rapidly in brilliance as they go outwards. This pattern represents the very best that any lens can achieve in re-imaging a luminous point; the possibility of obtaining a point image of a point object does not exist either in theory or practice, and the excellence or mediocrity of the target-like pattern depends on the relative size of the central disc and its brilliance with respect to the inner rings.

If the fine adjustment is moved slowly, so as to pass from a focal position above the artificial star to one below and back again, a characteristic series of changes will be seen in the pattern.

With a cover which is too thick, so that excessive spherical aberration is present, the star will change on upward focusing into a ring of light, and on focusing downwards into a disc with misty edges. The essential feature is that the appearances above and below the focal plane are dissimilar, indicating that the light is conforming to a caustic curve, with the oblique focus above the axial focus (Fig. 45A).

If the microscope tube is now shortened, or the correction collar adjusted for a thicker cover, a corresponding over-correction will be introduced in the lens system, which will counteract the excessive initial

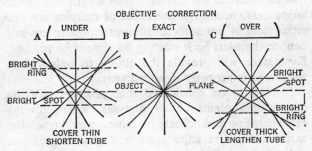

Fig. 45. *Appearance of the image of a luminous point in different conditions of objective correction.*

aberration. The best indication of the amount of adjustment needed is obtained by focusing in the "ring" position, and then adjusting the tube length or collar until the central focus is restored. The process is repeated until the setting is found at which the artificial star goes symmetrically out of focus in each direction (Fig. 45B); this represents a condition of spherical compensation in which the image is most clearly defined, with the smallest central disc and dimmest rings.

Of course, if the cover is too thin, the process is reversed; the ring of light appears below the star and the disc above (Fig. 45C), and the tube has to be lengthened to balance the appearances.

With a star-like object, the correction is quite easy to see, but in practice it is more usual to have to deal with a feature darker than its surroundings. This is much more difficult, as the asymmetry of the out-of-focus appearances is harder to assess, and utterly opaque points are not very common. Most particles tend to act as lenses, transmitting some light and redistributing it in the process, so that the effect of any adjustment is masked. In these circumstances it is best to maintain the standard tube length and collar adjustment, because if the effect of carrying out the adjustment is not sufficiently appreciable to be evident to the user, he is wasting his time if he persists. Experience will provide the ability to judge when the image is properly in adjustment.

It used to be considered that practice with the microscopic plants called **diatoms,** which display extremely fine and delicate markings, would teach the value of objective adjustment, and also illumination, in the best possible manner, but this is not altogether true. The structure of the more "difficult" diatoms is ambiguous where light microscopy is concerned, and the image seen depends on the degree of out-of-correction which is employed; in a state of perfect correction, the image becomes invisible.

The correction of spherical aberration and the fulfilment of the sine condition are described above as though applied only to rays of one specific frequency. This circumstance is met where truly monochromatic light is used, as in ultra-violet work where the individual spectral bands of sparks are used, but it is most unusual. The ordinary objective has to be capable of utilizing light of different colours simultaneously, so that coloured specimens may be seen effectively. This imposes another task on the designer.

Chromatic Aberration

The refractive index of a substance is not equal for all frequencies, but is higher for the higher ones, corresponding to the blue end of the spectrum (Fig. 46); the proportion in which the refractive indices

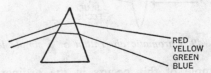

RED
YELLOW
GREEN
BLUE

Fig. 46. Dispersion of white light by a prism.

differ is known as the **dispersion** of the substance. This varies from one substance to another (Fig. 47),

A   A

GLASS I   GLASS II

Fig. 47. Different glasses show differences in dispersion.

and also in different parts of the spectrum, one type of glass being more dispersive at the blue end, and another at the red, for example (Fig. 48).

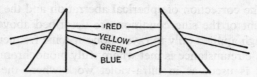

Fig. 48. Local differences in dispersion of various colours.

A great part of the difficulty in the design of microscope lenses arises from these twin properties. The effect of dispersion is to cause a simple glass lens to split up white light into its component colours, and to bring each to a slightly different focus (Fig. 49).

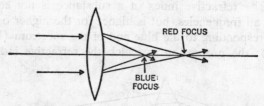

Fig. 49. Chromatic aberration of a simple lens.

An example of this occurs in the ordinary Abbe Illuminator; if the microscope lamp is focused on a bare slide, and a small light-source such as the filament of an electric lamp is brought into focus by the condenser, it will be found possible to change the colour of the image simply by slightly raising or lowering the condenser.

This effect is called **chromatic aberration,** and its correction was the advance which turned the compound microscope from a toy into an instrument early in the nineteenth century. The fact that different glasses differ in dispersion independently of refractive index—a point which escaped Newton, who looked into the matter over a century earlier—enables lenses to be made of two components having opposite and balanced chromatic errors (Fig. 50).

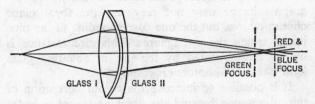

Fig. 50. *Achromatic correction.*

Achromatism and Apochromatism

Owing to the irrationality of the spectrum—the differential degree of dispersion in different parts of the spectrum mentioned above—the correction is not, however, perfect. Microscope lenses are normally corrected for the most visually effective colour, apple green, and the other colours are paired off, red with blue, orange with green, etc. This produces a result quite acceptable in ordinary use, giving a substantially colourless picture; such lenses are said to be **achromatic.**

Very minute colourless bodies, however, appear greenish in colour, and this may lead to a mistaken belief that the colour is genuine and not what is termed an **optical artefact**—in other words, a spurious appearance produced by the conditions of observation. This is an example of the type of error mentioned earlier as likely to arise unless the limitations of the microscope are appreciated; small transparent greenish bodies may well exist, but the fact that such are visible cannot be taken as *ipso facto* proof that they are present without confirmation.

A difficulty arises also with achromatic objectives when they are used for photography. The emulsion of the ordinary photographic plate is more sensitive to blue than to apple-green light, so that the photographic image may not represent the focal plane chosen by eye, but the one corresponding to the blue image which the eye ignores. This circumstance is largely circumvented by the use of coloured glass light filters for photomicrography.

It is possible to improve the colour correction of the microscope beyond the combination of pairs of colours. Objective design in the mid-nineteenth century was largely empirical, though some excellent lenses were produced. At this time the physicist Ernst Abbe became scientific adviser to the firm of Carl Zeiss, which had lately become interested in microscopes. Abbe brought a scientific attitude to his work, and in a series of papers published in the early eighties he discussed microscopic vision, the action of objectives, possible ways of improving them, and illumination. Then in 1886 he produced a new type of objective which instantly killed all others stone dead as far as research was concerned. These new ones yielded images free from colour, brighter than hitherto, and with increased resolution, which made it possible to increase the total magnification employed.

This pattern was designated **apochromatic,** and was characterized by the use of the mineral fluorite to eliminate the residual colour shown by the achromats; instead of being corrected for two colours, they were corrected for three. This was achieved at the

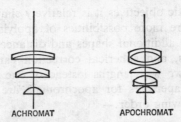

Fig. 51. Construction of apochromatic objective compared with equivalent achromatic objective.

cost of greater complexity (Fig. 51), and part of the correction was obtained not by the objective components but by the eyepiece, which had to be constructed to neutralize differences of size between the coloured components of the image (Fig. 52). An

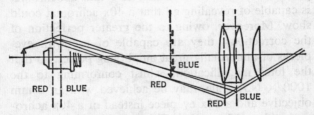

Fig. 52. Apochromatic objective, primary images and compensating eyepiece which eliminates chromatic differences of magnification.

apochromatic system, therefore, consists of an **apochromatic objective** and a **compensating eyepiece** designed to match.

Apochromatic objectives can be used for photography without colour screens, but their superb freedom from colour is the least of their advantages. It is possible to construct achromats with greater numerical apertures than those listed on page 70, and this has frequently been done, but in the case of

apochromatic objectives it is relatively simpler to do
so; there are more possibilities of applying corrections in the additional shapes and distances between
components, and spherical correction can be provided for two wavelengths instead of one only. The
customary apertures for apochromats are therefore
of the following order:—

Focal Length mm	Magnification	N.A.
16	10x	0.3
8	20x	0.65
4	40x	0.95
3	60x	1.3 or 1.4
2	90x	1.3 or 1.4

It will be seen at once that as far as resolving
power is concerned, a 12 mm or 8 mm apochromat
is capable of revealing all that a 40x achromat could
show. Moreover, owing to the greater perfection of
the corrections, they are capable of use with eyepieces of abnormally great amplifying power, so that
the total magnification, whilst conforming to the
1000 × N.A. rule, may be achieved with an 8 mm
objective and a 18x eyepiece instead of a 40x achromat with a 10x ocular. This provides a wider field
of view, but the depth of focus is no greater with
the apochromatic objective, as the numerical apertures are equal, so that the fine adjustment will be
needed to focus it.

It is generally agreed that if a microscopist were
to be offered the choice of one single objective, he
would select the 20x apochromat as having the greatest versatility of any. Many microscope makers now
offer a series of objectives designated fluorites or
semi-apochromats. Spherical and chromatic corrections for two wavelengths are provided.

Chapter Six

LIGHT AS AN IMAGE-FORMING AGENT

Normal Circumstances

IT was remarked earlier that the microscopist practically manufactures the image he observes by the steps he takes to illuminate the specimen. The adjustment of the body and its lenses leaves little scope for variation; it is either efficient or inefficient to a greater or lesser extent. The illumination, however, can usefully be performed in several ways, which produce different effects capable of accentuating different features of the specimen. For this reason it is important to look on the final image not as a faithful picture of the object, but as a demonstration of the manner in which it affects light in the circumstances selected.

In ordinary life, we rely on a general diffuse lighting to allow us to see objects; light is reflected to our eyes by the surfaces of the objects we see, and provided that enough is available, we see clearly. Where objects have a polished surface it may be impossible to discern their exact shape in certain circumstances, and if the light is not diffused but in the form of a beam it may give rise to misleading effects, though the fact that we have two eyes, which each see a slightly different view of the object, is of great assistance in such cases. The eyes are infinitely experienced in interpreting natural views, and make sense of the situation by subconscious adjustments of focus and convergence, and whilst instrumentalists are apt to be contemptuous of their efficiency, they unfailingly provide adequate images in circumstances in which a refined instrument would simply fail to function at all.

We arrange anything we wish to examine so that the light falls on it from behind us; only stained glass, wine, and other transparent objects of which the color or clarity is the sole point of interest are viewed against the light.

When an object is examined under the microscope, however, conditions are quite different from those with which we are familiar, though the distinction between incident and transmitted illumination can still be recognized.

It is possible to recognize six different and independent ways in which the object can affect light, and these can be deliberately used to assist the appreciation of the object by a competent microscopist, though one who is badly instructed may be misled by them. Quite simply, and without going into physics, they may be listed thus:—

1. Absorption. The most obvious indication that something is present in the field of view is the obstruction of some or all of the light when it intervenes between the eye and the light-source (Fig. 53). This is the commonest condition in microscopy, and the most uncommon in normal experience, as has been pointed out. In spite of the unfamiliarity, however, it is easy to understand suitable objects when examined by what is called **transmitted illumination;** the boundaries of particular parts of the object are marked by differences of colour, and the thicknesses by variations in brightness.

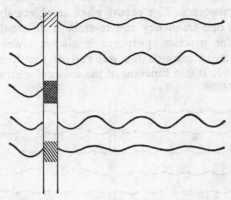

*Fig. 53. Light passing through an absorbing object is re-
duced in intensity.*

A great many natural objects are well-adapted to
this form of examination, and those which are transparent can usually be made to absorb light by being
dyed. This process of dyeing, termed **staining,** is of
the very greatest importance in microscopy. The action of many stains on a tissue is not uniform, as in
textile dyeing, but is controlled by the biochemical
nature of the object, so that various substances can
readily be differentiated from each other, and the
structure of an initially transparent section of an or-
gan displayed in diagrammatic fashion as individual
cells, nerves, glands, blood-vessels, and possibly parasites.

The means by which this result can be achieved
are very numerous and specialized, and every particular group of specimens has a standardized régime
by which a conventional display of its attributes can
be obtained. The methods are constantly being extended, and form a specialized study which requires
constant attention if its progress is to be followed.

2. Fluorescence. This occurs when an object absorbs light of one frequency and re-emits a different one, which for practical purposes is always lower—that is, displaced towards the red end of the spectrum (Fig. 54). It is a function of the molecular structure

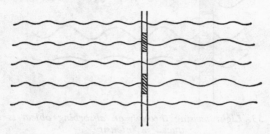

Fig. 54. Fluorescence introduces a change of frequency and thus of colour into the illuminating rays.

of some component of the specimen, which may undergo chemical change as a result of the energy obtained from the illuminating beam, and it may be compared in a loose analogy with the jangle of a sonorous object in response to an appropriate musical note.

Certain natural objects, of which fluorspar is the most famous, glow under the influence of ultraviolet light, and there is a range of chemical reagents which respond to the stimulation of characteristic wavelengths. The luminescent colours of posters are derived from some of these, and several can be employed to produce a selective effect in specimens, so that if appropriately illuminated they can be used to distinguish the structures which have absorbed them.

From the technical point of view, these **fluorochromes** are in exactly the same category as ordinary stains in that the correct one has to be selected and applied in the specific conditions of pH and concentration which lead to its absorption by the required material. Optically the case is of great interest because it provides the one circumstance in which the theoretical condition of a self-luminous object is fulfilled, and might thus be expected to form an exception to the practically ubiquitous role of diffraction by the object as the principal image-forming attribute of the light-specimen system.

Fluorescence microscopy has been greatly advanced in recent years, and unquestionably has a very important future. In the past it suffered from the requirement of ultra-violet illumination, which demands special equipment, but recent developments have shown that by using suitable colours for illumination, with standard equipment, it can be applied without great inconvenience, and this will unquestionably lead to its increasing use.

Fluorescence may be considered as the opposite of absorption, since the latter reduces the range of wavelengths passing the object, whilst the former introduces new ones. Specimens are examined for fluorescent effects on a dark field, in order to show up the colours, which are frequently very brilliant when extraneous light is excluded from the image.

3. Reflection is not a property of microscopical specimens commonly exploited except in the field of metallurgy, in which it plays a predominant role (Fig. 55). It occurs at surfaces where there is a con-

Fig. 55. Reflection from smooth and rough surfaces.

siderable difference of refractive index between the two materials in contact, as for example at the face of a crystal or metal specimen, or at an internal boundary such as a cleavage plane or a bubble, particularly when the light falls obliquely on the boundary.

The apparatus used to obtain reflective effects is specialized, and depends usually on **incident illumination,** in which light is concentrated onto the upper surface of the specimen, which is left uncovered in order to avoid reflection from the cover.

Incident illumination can vary from **vertical illumination,** the standard method in metallography, in which light is projected down through the objective, which acts as its own condenser, to **oblique illumination,** in which a beam of light is directed downwards at any angle between the vertical and a grazing incidence. It can be symmetrical, when the light comes from all azimuths, or lateral, revealing inequalities of the surface by their shadows and thus permitting an estimation of their height.

The apparatus used for this method of illumination, apart from lateral oblique lighting, which can be obtained by simple means, is specialized, and is not commonly available in biological laboratories, though its possibilities in this sphere are once again becoming recognized. Until the end of the last century a biological microscope was normally equipped with apparatus for oblique incident illumination under low powers, but the standardized method of mounting specimens so as to avoid differences of refractive index, which interfere with an absorption image, has led to its exclusion from general use.

It may be considered that reflection is the basis of the **dark-field** method of illumination, in which a bright image is produced on an unlit field, but it is on the whole easier to consider this as a special case of image formation by diffraction, though the distinction is somewhat academic.

If a transparent specimen is illuminated from below by light which is too oblique to pass directly into the objective, so that the empty field of view is dark, features of the object which show abrupt changes of refractive index will appear self-luminous on the black background (Fig. 56). This provides a very searching method of illumination, which will reveal particles which are, like stars, too small to be distinctly seen, but indicate their presence by a circular diffraction pattern. Except with low powers it makes considerable demands on the quality of the

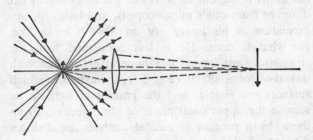

Fig. 56. Dark-field illumination by reflection.

optical system, including the slide and coverslip; its nature will be explained in the discussion of diffraction, and its application in the section on practical illumination.

4. Refraction is the property of the specimen which causes light to be deflected from its straight path owing to variations in its speed of transmission through different media. The whole action of the mi-

croscope depends on refraction as far as the lenses
are concerned, but refraction in the specimen is a
condition which interferes with other image-forming
properties, and is, therefore, avoided in the ordinary
way. The user of stained specimens, who can look
at his specimen as though it were a stained glass
window, gains no help from differences in refractive
index, which affect his image much as a rising cur-
rent of hot air would the view of the window, and
accordingly simplifies the interpretation of this image
by eliminating refractive differences as far as possible
by the use of mounting media having about the same
refractive index as protoplasm or glass.

Only the petrologist, examining thin sections of
minerals, makes deliberate use of refraction as an
aid in his work, but he works in a manner somewhat
different from other microscopists, and bases his appreciation of his image not on what it looks like,
but what it does. He is less concerned with the
straightforward image of a mineral grain than with
its out-of-focus effects (Fig. 57), the extent to which
surfaces are visible, and the pattern of light to be
seen in the upper focal plane of the objective. From
these, by a process of mental analysis, he deduces
the optical characteristics of the components of his

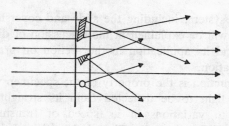

Fig. 57. Refraction from two crystals and a bubble.

specimen, and thus identifies the minerals and hence the rock which they constitute.

This is probably the most intellectual form of light microscopy, and demands a proper understanding of optical crystallography; it is an unsurpassed training for any type of microscopical work, as the petrologist is unlikely to fall into the error of accepting any image as a definitive picture of a specimen—he knows how such an appearance could be produced.

The problem of the examination of living biological specimens which show slight variations in refractive index, but no other property which could be used for image formation has been recognized for many years. It has largely been solved by the introduction of interferometric methods, which register in visible effects the relative delay suffered by light beams which traverse different parts of the specimen. These methods, which are based on the deliberate violation of the classical concepts of proper image-formation, will be described later.

The properties so far described—absorption, fluorescence, reflection and refraction—can all be described in simple terms without calling on the wave theory of light. If light were not a kind of wave action it would still be possible for them to exist. The two remaining properties, however, **polarization** and **diffraction** both depend entirely on the wave theory, and cannot be described in other terms.

Polarization may be loosely described as the coming-out of light waves so that the vibrations of the individual rays in a beam all occur in the same plane. In ordinary light originating from a single point source the individual waves are all of equal frequency and in the same phase, and will react and interfere with each other in suitable circumstances. Different points in the source each produce their own uniquely coherent waves, so that the total is an aggregation of independent waves each having its own distinctive frequency and phase condition and plane of vibration—only the velocity of the rays is common to all.

The behaviour of the beam as a whole, or of the rays themselves, is entirely symmetrical, as individual idiosyncrasies are cancelled out statistically. If, however, the vibrations are restricted to a single plane, the rays have distinctive properties in the plane of vibration and in the one perpendicular to it. These are familiar from the use of polarizing anti-dazzle glasses to quench reflections.

If a beam of ordinary light falls on a transparent surface, it is split into two components, one of which is reflected and one transmitted (Fig. 58). These are polarized in mutually perpendicular planes, transverse and perpendicular to the glass surface, though

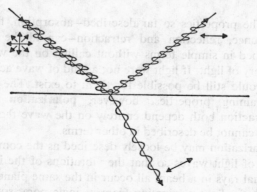

Fig. 58. A beam of ordinary light polarized in perpendicular planes by a transparent surface.

the effect is not complete; it reaches its maximum when the refracted and the reflected components diverge at right-angles.

Conversely, if a beam of polarized light falls obliquely on a transparent surface, its behaviour depends on the orientation of the light waves with respect to the plane of the surface. If the vibrations are perpendicular to the surface, the beam will penetrate it and enter the material behind (Fig. 59A); if they are transverse, the whole beam will be reflected from the surface (Fig. 59B), the vibration directions in both cases remaining unchanged.

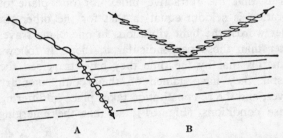

Fig. 59. Distinctive behaviour of perpendicularly-polarized beams falling on a transparent surface.

It is a property of crystals, which essentially have an orderly lattice structure of molecules, that light waves traversing the crystal can only vibrate in fixed planes, at right-angles to each other (Fig. 60). A

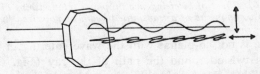

Fig. 60. A crystal transmits light vibrations only in perpendicular planes.

beam of ordinary light falling on the crystal is re-organized into two sets of waves vibrating in perpendicular planes, which pass independently through the crystal and join again after leaving it. The vibrations can be considered as quite random initially, but those oblique to the two planes of transmission are divided into appropriate components in these two planes, and re-combine in an appropriate oblique orientation afterwards.

The "appropriate oblique orientation" is not necessarily that in which they entered the crystal, however; a feature of the transmission of light by crystals is that the refractive index for one plane of vibration is seldom equal to that for the other. In other words, the light vibrations in one plane travel faster than those perpendicular to them. It follows from this that by the time they have traversed the crystal, they have gained on their complementary waves, and the re-constitution takes place in differing phase conditions (Fig. 61), so that the emerging

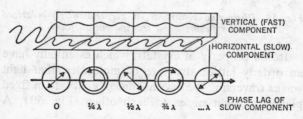

Fig. 61. Effect of birefringence on phase relationships of the two components of a transmitted beam of polarized light.

wave is not co-planar with the wave entering; it has been twisted round the path of the ray (Fig. 62).

With ordinary light this does not produce any visible effect, as the eye is not capable of detecting dif-

Fig. 62. Rotation of plane of polarization by a crystal.

ferences of phase in any case, whilst the total effect of the action of the crystal on all the rays present would produce statistically the same distribution of wave-planes as the original.

When polarized light is used, however, the result is readily visible. It is convenient to polarize ordinary light by passing it through a **polarizing filter,** or **polar,** and the effectiveness of this process is demonstrated by passing it through a second similar polar (Fig. 63). If this is rotated about the optic axis, it

Fig. 63. Ordinary light passed successively through two polarizing filters.

will be found that light alternately passes through the second polar and is obscured by it twice in a revolution (Fig. 64).

Fig. 64. Changes in intensity of light passing successively through two polarizing filters as one is rotated.

If two polars are arranged in the "crossed" position, which extinguishes transmitted light, and a piece of crystalline material (mica is convenient, and also the synthetic films of cellophane or polythene) placed between them, it will be found that if the crystalline sheet is rotated, it becomes alternately bright and dark when viewed through the upper polar, four times in a revolution, in perpendicular azimuths (Fig. 65). The dark positions are those in

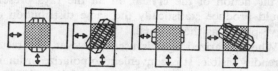

Fig. 65. Changes in light transmission as a crystal is rotated between two "crossed" polar filters.

which it does not affect the existing polarization of the light reaching it; the plane vibrations are passing neatly in the appropriate vibration direction, and emerging unaltered. In the intermediate positions, however, coloured light passes the second polar, growing steadily more intense as the crystal is turned from the dark position through 45°, and then dimmer again until complete extinction is reached.

The increase and decrease in intensity is readily explained by considering the proportion of the total light transmitted in each vibration direction; at 45° this will be equal, and any de-phasing will have its maximum effect at emergence in producing a variation in the original polarization plane.

Polarization Colours

The colour is not so obviously explained, but is due to the same cause. White light consists of a series of graduated wavelengths; these are rotated individually by the crystal into the plane in which they are extinguished by the second polar. With a thin crystal, only the shorter wavelengths may be in this condition, and with a thicker crystal, the longer waves also (Fig. 66).

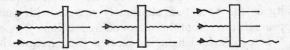

Fig. 66. Self-extinction of specific wavelengths by passage through crystal plates.

A continuous series of seven coloured "orders" named after Sir Isaac Newton, who first observed their relationships, is recognized in polarization colours, which differ from spectral colours in not being isolated pure entities, but subtractive colours— white lacking red, white lacking green, lacking blue, etc. The repetition of the series with slight changes represents multiple subtractions, with both red and blue interfering, or yellow and violet and green, as the wavelengths individually cancel themselves concurrently with others.

Confining attention to the strictly qualitative, or image-forming aspects, of this phenomenon, it follows that a group of crystals which is colourless in ordinary light will show different extinction directions and different polarization colours which enable them to be distinguished quite independently of other features arising from their shape or refractive differences or decomposition products or other attributes.

Occasionally birefringence may be combined with a strong absorption in one transmission plane. Crystals which exhibit this effect—**pleochroism**—vary in depth of colour or even hue if examined without an analysing polar whilst the lower polar is rotated. A different effect occurs where the two refractive indices in the plane of the specimen are widely different; the stock example is calcite, which twinkles as the lower polar is rotated with the upper removed. In this case the effect arises from pronounced differences in visibility as the specimen alternately merges with and appears distinct from the surrounding medium.

Where polarized light is reflected from irregularities or internal features in the specimen it becomes partly disorganized, so that the upper polar fails entirely to extinguish it. The effect is shown where micro-crystals occur in a glassy matrix, or where decomposition products or other alterations occur in crystal aggregates, and produces the appearance of irregular illumination of an otherwise dark specimen between crossed polars. Stresses in an isotropic medium cause double refraction, and become visible in similar circumstances as grey streaks or bright crosses.

This is the legitimate end of a consideration of polarization as a means of image formation, but it is only preliminary to a study of the exploitation of the phenomenon, which can be usefully employed in microscopy at two quite distinct levels, qualitative and quantitative. The books of the older school of microscopists all refer to the beautiful colours which can be produced in colourless chemical crystals and many other specimens, such as fibres, by its elementary application, and it has long been a favourite method for exhibition purposes, as it provides a most attractive and brilliant image.

The more advanced application of polarization phenomena depends on the determination of the optical constants of the specimens, and although a microscope is required for this, the instrument is only functioning casually as a microscope; its chief use is that of a measuring instrument and interferometer applied to small objects.

A real discussion of this field of work would involve a detailed initial consideration of crystal optics, and this would essentially be of somewhat restricted interest. The modifications incorporated in the design of the instrument to facilitate such work will be described in due course, and references to polarization microscopy will be found in the bibliography.

Attention is sometimes drawn to an improvement in resolution obtained by the use of an analyser without a polarizer in the examination of finely-marked specimens such as diatoms. The explanation of this appears to be the partial polarization of diffracted rays according to their orientation, which enables a slight difference in the relative visibility of different parts of a composite pattern to be produced. The effect is therefore, not strictly a gain in the resolving power, but an improvement in visibility owing to the control of glare, enabling resolution to be recognized where it might normally be masked by stray light.

Diffraction

The final process by which images are formed—diffraction—is by far the most important of the six. It has been left until last to permit more detailed treatment, as the whole practice of microscopic examination depends largely on diffraction, and the methods of illumination commonly used are intended to exploit this phenomenon.

Diffraction may be defined as a systematic scattering which occurs at the sides of light beams. This takes place wherever light passes across an edge (Fig. 67), and causes it to be deflected into the shadow,

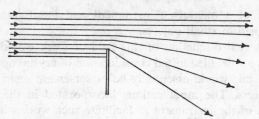

Fig. 67. Diffraction at an edge.

which therefore, never has a really sharp margin. It is most readily apparent when light traverses a narrow aperture, and the size of the beam is mechanically limited by the edges. In these circumstances the light rays, which normally travel in straight lines, tend to fan out slightly after passing the aperture. In consequence, it is possible to see the light behind the aperture from a position out of line with the aperture and light source.

The extent of the diffraction, which is usually masked by the much greater intensity of the main beam, is accentuated when caused by a narrow slit, so that there is a greater proportion of diffracted to direct light. In these circumstances the diffraction fan may be seen to extend through a considerable angle on either side of the direct ray, the intensity falling rapidly with increasing deflection (Fig. 68).

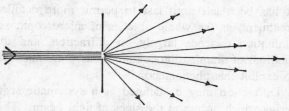

Fig. 68. Diffraction at a slit.
[Lengths of arrows indicate intensity.]

This bending is a function of the wavelength of the light; the longer waves are deflected more than the shorter ones, and indeed when the very short X-rays are used, the diffraction is so reduced that useful magnified images can be formed by shadow-casting.

It is worth noticing that a small opaque point which would cause light to be diffracted round its margin may thus be distinguishable in blue light, with a relatively short wavelength, when in red light the greater degree of diffraction may mask it entirely (Fig. 69). Where the diffracted rays occur in a me-

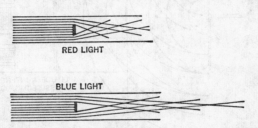

RED LIGHT

BLUE LIGHT

Fig. 69. Diffraction is less with shorter wavelengths.

dium of high refractive index—i.e. where light is travelling relatively slowly, and the wavelength is correspondingly reduced—the scatter is less than when it occurs in air, and resolution is better.

If two similar slits exist closely side by side in an opaque screen, and they are illuminated by coherent light, the resulting diffracted waves will be capable of interfering. Each slit will give rise to a diffraction fan, and these will be superimposed (Fig. 70). In these circumstances a section across the field parallel with the plane of the two slits will encounter waves in additive and in subtractive phases, which will result in a series of alternate bright and dark bands parallel with the slits.

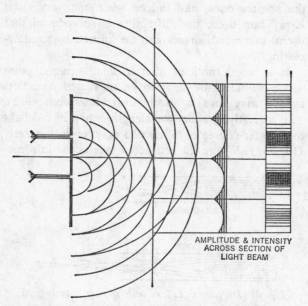

AMPLITUDE & INTENSITY ACROSS SECTION OF LIGHT BEAM

Fig. 70. Production of interference fringes by coherent light diffracted from two parallel slits.

The centre band of the striped field, opposite the mid-point of the bar separating the slits, will always be bright, as it is equidistant from each slit, and the waves will therefore arrive in phase and reinforce each other. On either side of this position, however, there is one which is half a wavelength nearer one slit than the other. Waves from the two slits will cancel each other out in these positions, where there will be dark bands. Outside these again are a pair of positions in which the difference in light-path is a full

wavelength, so that a condition for reinforcement
occurs, and similarly dark and bright bands follow
outwards from the centre of the field, the bright ones
having light-paths differing by a whole number of
wavelengths and the dark ones by an odd number of
half-wavelengths. It is easy to see that given the distance from the plane of the slits and the separation
of the slits and dark bands, it is possible to determine the wavelength of the light by pure geometry.

If the light used is white, consisting of a series of
different wavelengths, the various colours will be extinguished individually in order, resulting in a series
of Newtonian coloured bands, and these coloured
bands become superimposed rhythmically as the deflection from the centre increases, with the result that
a series of pure spectra occurs on either side of the
undeviated ray, running from blue to red outwards
(Fig. 71).

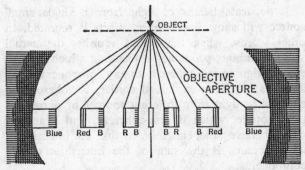

*Fig. 71. Diagrammatic appearance of diffraction spectra seen
in the objective aperture and arising from a striped object.*

The angular displacement of these successive
spectra depends on the interval between the generating slits and on the refractive index of the medium

in which the spectra are formed. The closer the slits the greater the angle, and the higher the refractive index, the less the angle.

It was the great optician Ernst Abbe of Jena, the scientific genius responsible for the high reputation of the Zeiss firm, who first appreciated the importance of diffraction by the object in image formation. He made this the basis of his outstandingly successful objective designs, and although his Diffraction Theory must now be regarded as dependent upon a simplification of the situation, it continues to be of service to microscopists because it provides a sound basis for explaining the phenomena encountered in interference microscopy and phase contrast, and for practical microscope manipulation in any conditions.

In a simplified form it may be comprehended without a mathematical description, if considered in stages:—

1. Separate beams of light from a single small source will show interference effects if reunited. In other words, when the beams reunite, the actual waves combine, producing a resultant which may be either brighter or dimmer than any component. Such light, as has been explained earlier, is coherent, in contrast to light from different sources or parts of a large source, which is non-coherent, and beams of which behave independently, so that the intensity of united beams is the sum of the intensities of the components.

2. Light which encounters an object is diffracted by its margins, and its internal structure if it is translucent. This results in a series of diffraction fans corresponding to the diffracting features, each fan consisting of a coherent set of diffracted rays, diverging

from the diminished main beam traversing the feature generating them (Fig. 72).

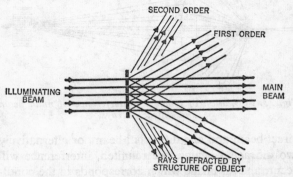

Fig. 72. Production of diffraction pattern by a striped object.

Thus, light leaving the object consists of two components—the direct beam, which behaves almost as though the specimen were not there, and the diffracted beams, which register in their direction, intensity, and phase, the minute structure of the object.

3. This dissociation of the original beam can be plainly demonstrated with narrow beams of light and very regular objects (Fig. 71), but in the general run of microscopic objects the structure is too irregular for the diffracted beams to be individually identifiable, although they retain their identities.

4. The groups of rays constituting the direct and diffracted beams which enter the objective separately are reunited in the image plane, where their interference produces the image of the corresponding feature (Fig. 73).

The formation of any interference image is therefore dependent on at least two components of the corresponding diffraction pattern reaching the image plane; otherwise there can be no interference. If the

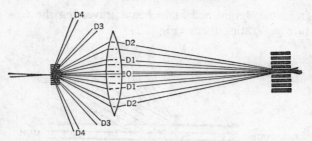

*Fig. 73. Formation of image by interference of main beam
and diffracted rays.*

direct beam and one diffracted beam, or alternatively
two diffracted beams, are reunited, interference will
occur, and an image which corresponds to the conditions of its formation will be produced.

The finer the diffracting structure, the greater the
angle between the direct ray and the first diffracted
ray, and those between the diffracted rays of ascending orders, and consequently the wider the cone of
acceptance of an objective which is required to collect them. This is the reason for the advantage to
be found in using objectives with high numerical
apertures; their ability to embrace a more widely-
spread diffraction pattern far transcends any photo-
metric advantage which they may possess over lenses
of moderate aperture. The intensities of the diffracted
rays decline rapidly as their obliquities increase, so
that the closer and more accessible ones make the
greatest contribution to the image, providing the general pattern of the structure, whilst the more oblique
ones sharpen the detail. They, as it were, dot the
i's and cross the t's without essentially altering the
result.

It will be evident that the image is less dependent
on what is produced by the object than on what is

transmitted to the image plane. To the extent that the "signals" generated by the object are lost, by not entering the objective, or are garbled by having their phase relationships disturbed, the image can be expected to differ from the object.

It is worth considering a simple case, in which a parallel beam of light passes through a specimen consisting simply of a uniform series of slots. In such a situation, all the diffracted rays will lie in a single plane fan, perpendicular to the plane of the slots, and diverging from the main beam and each other at the appropriate angle, which may be taken for convenience as 10° (corresponding to a grating with 17,354 slots to the inch). Ignoring for the moment all but the main beam and the pair of adjacent diffracted beams, it will be seen that these embrace a total angle of 20°. If the grating is examined in axial light with an objective of aperture 0.18, which corresponds to an angular aperture of 20° 44', the three beams will all enter, and will form an image of the grating (Fig. 74A). If, on the other hand, an objective of 0.17 N.A. is used, with an angular aperture of 19° 30', only the main beam will enter the objective, and an image cannot, therefore, be formed, as no interference can occur (Fig. 74B).

If the illuminating beam is tilted towards the position occupied by either of the adjacent diffracted beams, so that instead of falling at the centre of the objective it falls towards the margin, one of the adjacent diffracted beams, which will move with it, will enter the objective, and an image will be formed (Fig. 74C). This image will be that of a grating with 17,354 lines per inch, though the relative widths of bars and spaces will not be accurately delineated, and errors in the spacing of the lines of the

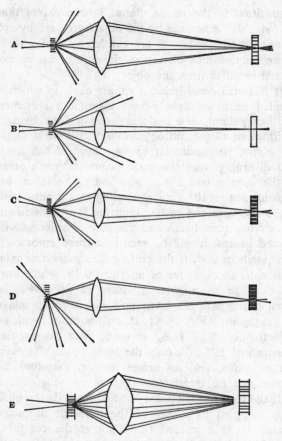

Fig. 74. A) Diffracted beams enter objective—image formed.
B) Diffracted beams do not enter objective—resolution fails.
C) Oblique illumination brings in one diffracted beam—
image formed. D) More oblique illumination leads to dark-
field image formed by two diffracted beams. E) Additional
N.A. collects more diffracted rays and results in a more
accurate image.

grating will be suppressed; the image will represent a first approximation to the object.

If the obliquity of the illumination is increased to the point at which the main beam is at the margin of the objective, the first diffracted beam will be almost central, and the second will be just outside the lens. In these conditions the objective could form an image of a grating almost doubly as fine as the existing one, in which the first diffracted beam would lie at the margin opposite the main beam, and both would be accepted simultaneously.

As the obliquity is increased still further, the main beam passes outside the objective at one side just as the second diffracted beam enters opposite, so that both first and second diffracted beams on one side of the fan are present together (Fig. 74D).

These two will form an image in which the relative widths of the slots and bars are correctly shown, but with their contrast reversed; the slots will be dark and the bars bright. If the objective had just enough aperture to embrace the main beam and these two together, the relative widths would have been correct and the contrast normal (Fig. 74E).

The importance of this lies in the twin facts that by directing the light obliquely through the grating, the resolving power of the objective can be made almost double what is possible in axial light, and that by excluding the direct beam a good image with a reversed contrast can be produced; this last is called **dark-field illumination.**

If by the use of special stops in the objective the main beam and second diffracted beam are admitted, but the first diffracted beam is stopped, the image registers the structure of an object which would give rise to such a distribution of diffracted rays—a grating twice as fine as the actual object (Fig. 75).

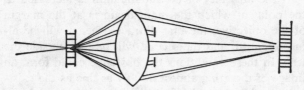

Fig. 75. Production of a spurious resolution by excluding important image-forming components and accepting others of less significance.

This was demonstrated by Abbe as evidence to support his theory of image-formation by diffraction, and caused considerable anxiety among microscopists until it was realized that such spurious images could only be produced by deliberate tricks, and would not arise unrecognized in normal circumstances.

When the processes of image-formation were imperfectly understood, oblique lighting with the mirror was employed for the purpose of obtaining maximum resolution, and a number of microscope designs were produced in which the entire substage assembly could be swung round to the side to obtain oblique light. This expedient is both clumsy and unnecessary, as a good substage condenser is more than adequate to fill the objective with light whilst in axial alignment with it.

A solid cone of light equal to the angular aperture of the objective may be regarded as the summation of all conditions of oblique lighting which the objective can accept (Fig. 76), and as the conditions are symmetrical about the optic axis, resolution is the same in all azimuths, which takes care of the

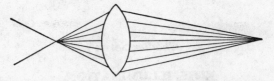

Fig. 76. Use of a wide cone of solid illumination represents the optimum condition for accurate image formation.

usual condition in which the object structure is not linear but complex.

If dark-field conditions are required, an illuminating cone of wider angle than the objective N.A. must be used, with the central portion corresponding with objective N.A. stopped out. In these circumstances the only rays entering the objective are those diffracted (or reflected) by the structure of the object.

From the foregoing remarks on diffraction, it will be appreciated that where resolution is important it is advantageous to use light of the shortest wavelength suitable—green for vision, blue or ultraviolet for photography—combined with the maximum working N.A. in the optical system, and to be particularly careful in the matter of objective correction.

Chapter Seven

THE ILLUMINATION
OF THE OBJECT—THE SUBSTAGE

Significance of Illumination Glare

It is true in general that the parts of the microscope above the stage cannot be wrongly used, but their efficient working can be prevented by wrongly adjusted illumination. This is so little recognized that many users of the microscope never obtain a good microscopic image, or even know what is possible in this respect. Almost invariably this is the result of failing to utilize adequate apparatus, and once the correct methods have been demonstrated, the improvement in work is radical and permanent.

The basis of the proper procedure is quite simple, and can be succinctly described as letting the objective do its job. The significance of numerical aperture in an objective has already been explained; the wider the cone of light which the objective can receive from the object and transmit to the image, the more exactly the object will be delineated. Now, except in the case of objects which scatter light noticeably by diffraction, the angle of the cone of light received by the objective is approximately equal to that of the cone of light reaching the object.

It follows that illumination of the object should be provided by a converging beam of light which comes to a focus in the specimen, forming a cone with an apical angle corresponding to the N.A. of the objective. The illuminating cone of light should not be larger than the cone of acceptance of the objective, and in fact it should always be slightly smaller. This is to avoid the phenomenon known as **glare,** which is haziness of the image due to the presence of dispersed light which increases the general brightness of the field and reduces contrast (Fig. 84).

The optimum relationship of the illuminating cone to the image-forming cone depends on several variables, so that it is necessary for the illuminating cone to be variable also, and this is arranged by constricting it below the specimen by means of an opaque screen having an aperture in it—the **substage diaphragm.**

In the very simplest case possible, this consists of a round hole which is raised and lowered below the specimen, thus controlling the angle of illumination available from an extensive light source such as the open sky (Fig. 77).

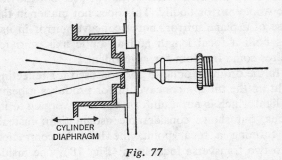

CYLINDER
DIAPHRAGM

Fig. 77

This simple construction suffers from the defect that it reduces the amount of light available at the specimen below the general level of illumination, whilst to produce an image of useful brilliance the specimen requires to be lit extra brightly. Apart from this, it requires a very extensive light source, and if this is restricted, as for example by the use of a small microscope mirror, the effect produced is not that of a converging cone of light, but of a parallel or diverging beam. This, of course, will come to a focus in the upper focal plane of the objective, which will thus be working at what amounts to a very small numerical aperture.

The Concave Mirror

In low power visual work, the traditional means of augmenting the illumination is the use of the concave mirror. This is a device of venerable antiquity, but its simplicity appears to have been partly responsible for defects in its application to the microscope, and in its use. The tendency nowadays is to mount the mirror in gimbals which allow only the freedom of altering the inclination of its surface, not moving the whole mirror bodily. This does not matter in the case of a plane mirror, but a concave mirror in use has both a focal length and an optic axis, and requires some freedom for adjustment.

In the ordinary oblique position used for reflecting light up the tube, a concave mirror produces a beam of light which is not a uniform cone converging to its focus, but shows considerable astigmatism; instead of forming a focal point, the light is contracted into two transverse focal lines (Fig. 78), one inside the mean focus and the other outside. In effect, the oblique mirror has two partial focal lengths, with a zone between the two foci where the light forms a disc. This is quite tolerable for visual use, but shows badly in a photograph, which is usually affected by uneven lighting as a result of unequal quantities of light reaching the specimen from different parts of the mirror which have been presented to the light source at different obliquities (Fig. 79). It is possible to avoid this effect if the mirror is capable of being adjusted so that its mean optic axis can be moved

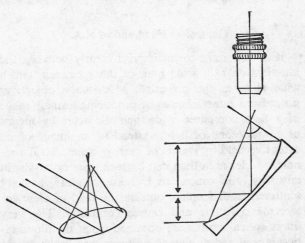

Fig. 78. Astigmatism of a concave mirror in working attitude.   *Fig. 79. Uneven illumination caused by mirror mounting.*

to balance differences in brilliance, but it is very doubtful whether it is worth doing so. A Victorian microscope with a double crank joint in the mirror-stem would allow an even field to be secured by careful adjustment, but in modern circumstances there appears to be every advantage in using only the plane side of the mirror, with a long-focus lens in the substage, as this eliminates the difficulties and provides a much better result.

The belief, occasionally met, that the concave mirror should be used to provide parallel light from a lamp for the condenser, has absolutely nothing to commend it. The only possible result of such action is to prejudice the action of the substage condenser by providing a beam of light which is astigmatic and inevitably oblique.

Control of Field and of N.A.

It is important to distinguish clearly between the illumination of a wide area of the specimen, and a wide N.A. in the objective. Microscope objectives differ from camera lenses by producing a small image of a field compared with their diameter by means of a wide cone of light, instead of an image of an extensive field by means of narrow cones. To illuminate a wider area than can be seen does not assist in any way, and causes an increase of glare, as light scattered from features outside the field of view enters the objective and causes confusion. The very finest optical results are obtained when the illumination is confined to a small uniform area in the centre of the field of view, but for general use, uniform illumination of the entire field of view, but no more, is best.

These separate requirements, the provision of a cone of light corresponding to the N.A. of the objective (Fig. 80), and of a powerfully illuminated area equal to the field of view (Fig. 81), are together satisfied by the use of the **substage condenser** below the specimen to produce an image of the source of light in the plane of the object (Fig. 82).

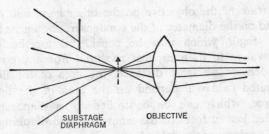

SUBSTAGE
DIAPHRAGM   OBJECTIVE

Fig. 80. Aperture diaphragm controlling N.A.

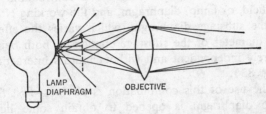

LAMP
DIAPHRAGM   OBJECTIVE

Fig. 81. Field diaphragm controlling area illuminated.

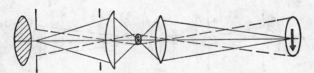

*Fig. 82. Use of a substage condenser to image field dia-
phragm on specimen.*

The objective then projects an image of the light
source, with the image of the object superimposed,
into the eyepiece. Considering the light source to be
a uniform luminous surface, it will be realized that
the extent of the field of view illuminated will depend on the focal length of the condenser and the
distance of the lamp, but will be independent of the
diameter of the condenser lenses. The illuminated

aperture of the objective on the other hand will depend on the diameter of the condenser, and on its focal length, which may be considered together as equivalent to its numerical aperture. For any given condenser and lamp distance, the area of the illuminated field will depend on the area of the light source, which can be controlled by a diaphragm placed just in front of the lamp—the **field diaphragm** —to conform to the field of view.

Thus the **area** of illumination is controllable by the field, or lamp, diaphragm, and the working N.A. by the substage diaphragm, which varies the effective diameter of the substage condenser, both being entirely independent and without mutual interaction (Fig. 83).

The use of this combination is obvious—first the lamp diaphragm is opened to obtain some light through the microscope, and the specimen is focused.

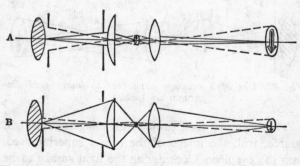

Fig. 83. A. Control of N.A. independently of field of view. B. Control of field of view independently of N.A.

Then the lamp diaphragm is closed, brought into focus by the substage condenser adjustment, centred by means of the mirror, and opened to the extent of the field of view. Lastly, the substage diaphragm is adjusted to the maximum aperture which can be employed without the specimen showing signs of haze, and the specimen examined.

The Nelson Three-Quarter Cone

The microscopist E. M. Nelson came to the conclusion that the best results were attainable when the condenser aperture was equal to three-quarters of the objective aperture, and that if this ratio were exceeded, the image deteriorated. This deterioration is due to glare, which may be defined as light which enters the image without contributing to its improvement. Stray light which merely produces an increase in the level of illumination may arise (Fig. 84) from

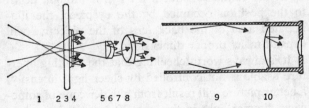

Fig. 84. Some causes of glare.
1 badly corrected condenser; 2 excessive area illuminated; 3 opalescent mounting material; 4 dirty or defective coverslip; 5 dirty objective; 6 internal reflections in objective; 7 cemented components separating; 8 excessive N.A. illuminated; 9 reflections in tube; 10 dirty or defective eyepiece.

reflections from the internal surfaces of the microscope, light diffused from imperfectly polished lens surfaces or from dirty or defective coverslips, or opalescent mounting media, or from the action of a bad substage condenser.

Nelson used an oil lamp, which was capable of lighting his complete field of view, and it was subsequently established by Conrad Beck that if the illumination were confined to a small area of the field, it was possible to work with a much wider illuminating cone, provided that this was not quite large enough to fill the objective completely; the margins of the objective lens ought to be left dark.

The extent of the objective which is illuminated can be judged by inspection down the tube with the eyepiece removed. It is necessary for the eye to be placed close to the end of the tube to obtain an accurate assessment; if the eye is held a foot or so above the open tube, the illuminated aperture will appear to be very small even with the diaphragm wide open, and this may cause misapprehensions about the quality of the condenser, but if the eye is brought down to the position occupied by the eyepiece, the illuminated area of the back lens of the objective will expand to its proper dimensions.

It is always worth checking that the limiting aperture usable is not controlled by sheer light intensity; a better picture will result from a wider cone of somewhat dimmer light in this case. It is most important that the substage diaphragm should not be used to regulate light intensity, although it has this incidental effect when properly used. Use of the iris as a dimming control is the besetting vice of the laboratory student, who is, however, often forced into it by the provision of an excessively bright bench lamp with no means of control. Working on prepared slides specially selected for demonstration, he rapidly acquires a firm conviction that the less the iris is open, the more he can distinguish, and passes on this gem of practical guidance to *his* hapless students, who all come to recognize that "Microscopy is bunk".

If the light is too bright for comfort, the reasonable course is to use a weaker source, or else to adopt some means of diminishing the intensity which does not interfere with the action of the microscope. A very simple expedient is to use a secondary source, such as a sheet of opal (**not** ground) glass in a holder lit from behind by a clear electric lamp. The intensity of the light diffusing from the opal screen depends on its distance from the lamp filament, and obeys the ordinary law of inverse squares, so that a slight adjustment of their separation has a considerable effect on the intensity. A set of plates with perforations of various sizes, mounted in front of the opal glass, enables the area to be controlled.

The Substage Condenser

From what has been said earlier about objective design, it will be appreciated that a condenser, if it is to satisfy the requirements of an objective completely, should be corrected to a comparable degree of excellence. A similar objective corrected for the thickness of the slide instead of that of the coverslip would be ideal, and this expedient has been used, but is both expensive and limited in scope, besides being unnecessary. It is not essential that the condenser should be capable of forming an image with the same precision as an objective, though many good ones approach this possibility. Its function is to provide a uniform cone of light for the objective in the first case, and only secondly to project an image of the source into the field of view. It follows that its most important feature is its spherical correction; ideally the whole aperture of the condenser should direct light from one point in the source to one in its image. If any part of the aperture is not directing light to the selected point, it is diverting it elsewhere, and thereby interfering with the uniformity of the illumination and disorganizing the light before the objective receives it.

Compared with the spherical correction, other features such as achromatism and focal length are subsidiary. Freedom from colour is always desirable, but for work with stained specimens or light filters it is not important. The focal length is significant in practice only in respect of the size of the field of view to be illuminated, as a short focal length produces a small image of the light source, so that a long focal length extends the range of usefulness of the condenser with various objectives.

From the designer's point of view, the focal length is a limiting factor. Given the postulate that the condenser must have the greatest possible **aplanatic aperture**—that is, the greatest N.A. over which spherical aberration is inappreciable—he has to strike a balance between spherical error and focal length. If a given design, providing a difference in focal length between the axial and marginal rays of x mm, is doubled in size to double the focal length, the focal error will also be doubled, making a value of $2x$ mm, whilst the total N.A. remains exactly the same as before. The aplanatic aperture, however, will now be less than with the smaller design.

Consequently, although the condenser is not restricted, like the objective, by limitations of diameter, it becomes more difficult to provide a high standard of correction in one of long focal length. British computers originally adopted the same standard size for their condensers as for objectives, so that both would fit the R.M.S. thread, but European practice was different. Originally, the European designers virtually ignored the desirability of correcting the condenser, which they regarded mainly as a means of collecting as much light as possible, and consequently adopted a wider diameter and longer focal length. In the course of time, both European and American designers have produced condensers of the highest excellence in the larger size, and the standard diameter specified by the Royal Microscopical Society for the sleeve fitting of the substage—1.527″ (38.8 mm) —is intended to accommodate any of them.[1]

The focal length of large condensers is typically 10 mm or more, and the back lens is about 25 mm diameter. They are made in both dry and oil-immersion patterns, with aplanatic apertures up to 0.95 and 1.4 respectively.

Condensers of the small type fitting the objective thread have similar aplanatic apertures, with focal lengths of 4 to 6 mm. For use with high power objectives, and particularly with immersion objectives, there is no advantage in a longer focal length, as the field of view is in any case very small, but the larger illuminated field provided by the larger condensers extends their utility to objectives of lower power, and they are, therefore, preferable for general use.

The importance of the substage condenser in microscopy is almost invariably considered from the point of view of resolving power. This depends on the effective N.A. of the objective, and, therefore, anything which improves the exploitation of the N.A.

[1] A 39.5 mm sleeve fitting is now widely employed as well as special bayonet mounts. 38.8 mm is used in the United States only by Bausch & Lomb.

available is advantageous. There is a formula of considerable age, though its authority remains obscure, which states that the working N.A. of a microscope is effectively

$$\frac{\text{N.A. of objective} + \text{N.A. of condenser}}{2}$$

In other words, the average of the two is the factor which governs resolving power. It is possible more easily to think of instances in which this formula is incorrect than of others where it is true, even bearing in mind that the condenser aperture must be lower than that of the objective if solid cones of light are used. It does, however, serve to link together the action of the two apertures, which is most important. We have seen that the direct rays from the object to the objective are accompanied by diffracted rays, feeble in intensity but essential to the formation of the image, and that these, because they diverge from the direct rays, will occupy a somewhat more obtuse cone than the direct rays.

The objective aperture is, therefore, fully utilized by the rays resulting from an illuminating cone of somewhat lower aperture, the relationship between the two depending on the dispersion of the diffracted rays, and hence on the colour of the light, on the refractive index of the mountant, and on the fineness of the structure producing the diffracted rays.

It has been tacitly assumed that as the aperture of the substage condenser is increased by opening the iris diaphragm, that of the objective is correspondingly illuminated. This is, of course, the theoretical sequence of events, but in practical circumstances it does not always follow.

Defects of Abbe Illuminator

The vast majority of substage condensers in use belong to the type called the **Abbe Illuminator,** which consists of two simple lenses, a wide lower component and a hemispherical upper with a combined focal length of about 11 mm. This is cheap to make, easy to use, and passes a lot of light, but it suffers from very severe spherical aberration; the axial focal length is some 2 mm longer than that of the outer zone. Consequently, although the N.A. of the marginal rays is 1.2 when it is used in immersion contact with the slide, this figure does not represent its practical N.A., as a point on its optic axis is illuminated only by an axial beam and the particular conical beam which cuts the axis at the point in question (Fig. 85).

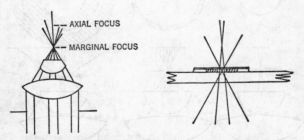

Fig. 85. Variation of focus of an uncorrected condenser with aperture, and pattern of illumination of a specimen.

This can be readily seen if the Abbe Illuminator is used with objectives of increasing power, the light source being adjusted to illuminate only the field of view. With objectives up to 0.6 N.A., the back lens of the objective can be quite evenly lit, though it will be found necessary to rack the condenser up to keep the lamp diaphragm in focus as the substage diaphragm is opened wider (Fig. 86A). It will be found that the image of the substage diaphragm visible in the upper focal plane of the objective vanishes as the diaphragm is opened; the objective can be filled with light by re-focusing the condenser whilst examining the objective, but the diaphragm is not itself very effective as the limit of the aperture is approached.

If now the object is replaced by one of 0.85 N.A., it becomes impossible by any manipulation of the diaphragm or of the condenser to fill the objective with light. A central disc corresponding to about 0.6

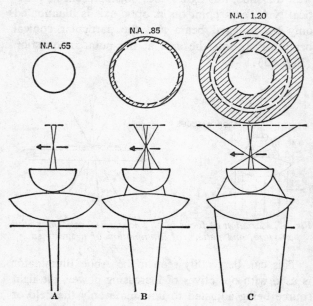

Fig. 86. Effect of lack of spherical correction in the condenser on the aperture of the objective.

N.A. can be filled (Fig. 86B), but beyond this stage the disc breaks up into a central area and a detached annulus, which can be arranged anywhere between the edge of the disc and the periphery of the objective lens by altering the focus. The same is, of course, the case if an objective of still higher aperture is used (Fig. 86C).

This condition, in which the objective is illuminated by a discontinuous pattern of light, is not suitable for refined work, because it is capable of producing fallacious appearances. The resolving power of the objective is interfered with, and the results obtainable cease to be predictable. Quite apart from these objections, the image suffers from uneven illumination, from colour differences between the central area and the margin which vary with the focal level, and from the virtual impossibility of obtaining a defined margin to the illuminated area; even at medium apertures the condenser is incapable of producing a distinct image of the field diaphragm.

If a large source of light is used, and the restriction of illumination to the field of view ignored, the image can be made fairly uniform, and the back lens of the objective appears evenly lit, but the result is obtained by what are really heterogeneous means; instead of putting the light where it is required, the uncorrected condenser is distributing it in a disadvantageous manner. All these defects lead to an increase in glare and diffuse light.

The Abbe Illuminator is adequate for use with a 40x objective of medium quality, with an N.A. of 0.75. Greater power demands a better condenser. There is nothing to be gained by using it in oil immersion except when it is employed as a dark-field illuminator.

When it is used with a patch stop, so that it projects a hollow cone of light (Fig. 87), the spheri-

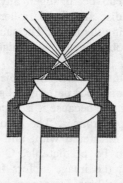

Fig. 87. Dark-field illumination by an uncorrected condenser.

cal aberration is not particularly apparent, as only a zone of the condenser is in use. Employed in this manner, it makes an excellent dark-field illuminator with objectives up to 0.65 N.A., the very errors in focus so obnoxious in ordinary use serving to provide an illuminated patch of uniform brilliance.

The Abbe Illuminator is made in a three-lens pattern with a stated N.A. of 1.4, and also with an aspheric lens surface which corrects the spherical aberration. Undoubtedly as manufacturing technique improves, the aplanatized Abbe will become more common, with advantages in every respect. At present the quality of Abbe-type illuminators is variable, and some are more carefully made than others.

The aberrations of the common type of Abbe Illuminator provide one reason for the difficulty experienced by many microscopists in trying to apply a tube-length correction in the case of ordinary specimens. Many objectives have a zonal spherical error, which varies in extent and sign between the centre and the margin of the lens, and if they are illuminated by axial-and-zonal rays with intrinsic spherical errors, a balance may be impossible to secure, and will inevitably be falsified when illumination is concentrated on another zone of the objective. The difficulty encountered in trying to make the correction in these conditions often leads to the suspicion that the whole procedure is somewhat fanciful, and an example of self-deception best ignored by a practical microscopist.

It must in fairness be pointed out that the Abbe Illuminator was originally designed in the light of Abbe's view that the most reliable microscopical image was one formed by a narrow cone of light; spherical aberration was not very severe in these conditions, as only a small portion of the lens was in use, though the diaphragm, by means of its ex-centring device and rotation (Fig. 89) could be made to provide a feather of light at any desired obliquity, and in any azimuth. With the optical means for accomplishing this intention there could be no reasonable quarrel, but the British school of microscopists at once repudiated the theory of the narrow cone on which it was based; practical experience had convinced them of the desirability of wide cones of light, and they had been using fully-corrected condensers for very many years.

The theoretical argument endured for a long time, and may be said to have ended in the favour of the wide-angle microscopists, but the great development of medical microscopy was taking place in the meantime in Europe, where a microscope condenser of any description had been practically unknown previously. It is said to have been Dr. Robert Koch, the great bacteriologist, who first recommended the substage condenser for medical studies, and its necessity rapidly became axiomatic. The immense prestige of the German scientists firmly attached the only condenser which they knew to every scientific microscope, where it was used in the way advocated by the British amateurs.

Abbe himself designed achromatic condensers capable of filling any objective with light, and it is an irony of fate that the name of the man who first explained the action of the objective, and designed the apochromats, should be inseparably welded to the uncorrected lens system which he insisted on calling an Illuminator, not a condenser. There seems to be every likelihood that it will prove to be his most enduring monument.

For use with low powers, particularly for projection, it is useful to have a very long focus condenser, of the type sometimes called a spectacle-lens condenser. As a rule substage condensers are too short in the focus for successful use with objectives of less than 10x magnification, even if their upper components are removed, leaving a combination with a focal length of 20–25 mm. This shows very noticeably in a photograph or in a projected image, particularly where the condenser used is the lower component of an Abbe Illuminator; in this case it is possible to select the colour of the image by focusing the condenser on the specimen, and it is very difficult to display a large specimen evenly illuminated.

There is a belief that a corrected low power condenser is not necessary, and at present few manufacturers provide them. One of the most favoured types in the past was Watson's Macro Illuminator, an achromatic aplanat with a focal length of 50 mm, capable of illuminating a specimen a centimetre across. It could be used either as a substage condenser, focusing the light source onto the specimen, or like the condenser of a lantern, placed close under the specimen and projecting the light into the objective; this is a favourite method when very large fields have to be evenly lit. This component was also popularly reputed to make the finest lamp condenser obtainable. Its manufacture has been discontinued, though low power condensers are still marketed by Charles Baker Ltd. When an unusual component such as this is required, it is well worth while enquiring of the various microscope makers if one is made; for some reason it is not uncommon to find that their catalogues list only the more obvious components, whilst the oddities which are eagerly sought second-hand are actually manufactured, but kept secret. It should be noted that Leitz and Zeiss Jena offer low power condensers for their macro objectives called spectacle lens condensers.

Mechanical Structure of Substage

The substage which supports the condenser requires a certain amount of consideration. It has to provide a focusing action in exact alignment with the optic axis of the microscope, and to accommodate a diaphragm, various condensers, and stops.

British and European ideas on the substage have been different hitherto. The British substage (Fig. 88), traditionally consists of a tubular sleeve $1.527''^2$ diameter, with focusing and transverse adjustments, to receive the diaphragm and condenser mounted together as a separate unit. The European substage (Fig. 89) is that originally designed by Abbe,

2 See note, p. 135.

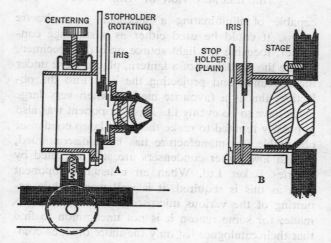

Fig. 88. A. The traditional British substage with centring action. B. The simple understage mount lacking a centring action.

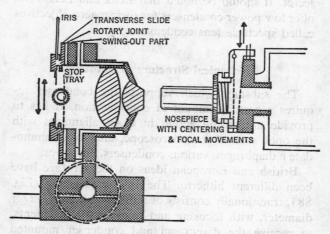

Fig. 89. The Abbe substage, with ex-centrable iris, and centring action applied to objective holder.

with a permanently centred tubular sleeve for the
condenser, and a swing-out plate below it which has
a rotating insert carrying an iris diaphragm. This is
adjustable radially, so that it can be located to provide oblique light from any position.[3] The European condenser mount is merely a short flanged tube
to locate the lenses in position. Now, however, the
British type is generally offered on U.S., European,
and Japanese microscopes, i.e. diaphragm and condenser as a separate unit with a swing-out filter
holder. The fittings vary.

Each system has many points in its favour. The
Abbe substage is convenient in use, but the objective
must be centred to the condenser, which defines the
optic axis of the microscope.

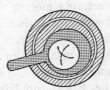

*Fig. 90. Centring mount for Zeiss cardioid dark-field il-
luminator.*

Interchangeability of condensers is secured by perfect workmanship, but it is naturally advisable to
centre the objectives to the most exacting condenser,
or the dark-field illuminator, and to rely on the
traditional indifference of the ordinary two-lens condenser to exact centration. Zeiss formerly produced
a mount for the cardioid dark-field illuminator to
centre it to the objective, and this can be used with
condensers of R.M.S. size. Instead of having radial
centring screws, it depended on the rotation of one
eccentrically bored tube inside another (Fig. 90),

[3] Nikon retains this feature.

and could not be used without adjustment on each occasion, because its orientation in the condenser sleeve was fortuitous.

Any centration applied to the condenser in this way puts the iris out of line to some extent, and though this might be taken up by the transverse rackwork, the spring catch marking the central position interferes with slight displacements. The proper way of centring an assembly of this pattern is by taking the iris aperture as the fixed centre, and centring the objectives to it with no condenser in place. Large condensers without centration are used as they stand, and rotated if necessary to minimize eccentricity, which is unlikely to be present, and when small ones are used in the centring mount, the levers are adjusted until the image of the iris is central. This plan eliminates the effects of differences between the centres of the microscope nosepiece and the iris.

European, American, and Japanese practice is now adopting the British application of centring screws to the entire substage, but the whole of microscope design is in a transitional state, and it is likely that the substage as it has been known in the past will be eliminated.

The British plan enables the optical centration of the condenser to be suited to that of the objective, which constitutes the fixed line of reference, without throwing the iris out of centre. Improvements in iris construction have practically eliminated the defects of eccentric opening and wandering centration which used to be prevalent, and it is usual with most makers now to supply both condenser and iris together as a unit, and to change both together.

Attempts to use a permanent iris in the substage, European fashion, have been made by British designers. The pre-war Watson Research Substage (Fig. 91) carried the plan to its logical end by using

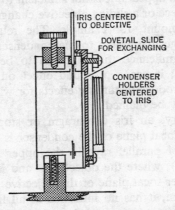

Fig. 91. The Watson Research Substage with centring condenser changers (c. 1930).

condensers mounted in individually centrable slides on an iris carried in the centring substage. Somewhat similar arrangements were offered by Zeiss and by Bausch & Lomb, though in the former the iris was not centrable.

There is at present a tendency to make the permanent part of the substage a simple U-shaped plate, and to relegate the centring action to the interchangeable components, in the manner of the original Akehurst changer. This is supposed to promote quick interchangeability of condensers, as each is permanently centred, but it involves a considerable duplication of fittings, and in some designs is a hindrance, not an assistance to convenient interchange.

Location of Iris Diaphragm

In fact, the interchange of condensers is not as simple as it appears; it is not sufficient to use a rotating nosepiece or a sliding objective changer on an old substage mount, for example. The relationship of the substage diaphragm to the condenser itself is of greater significance than appears to be commonly recognized; it has a profound effect on the image, and can produce disturbing effects for which the objective may be blamed.

In theory, an iris diaphragm or stop placed in the lower focal plane of the condenser is represented by a set of parallel rays in the upper focal plane (Fig. 94A), where the object lies, and is re-imaged in the upper focal plane of the objective. Located in this position, it has no influence at all in the plane of the specimen except to increase or decrease the angular aperture of the illumination; the uniformity of the illumination remains unaffected, though its intensity is naturally affected as the total available light is augmented or diminished.

In practice, there are objections to locating the diaphragm in this position; the more obvious are as follows:—

1. Most condensers have an internal lower focal plane.

2. The lower focal plane is always more or less curved, and an ordinary diaphragm could not be located in the curve.

It is, therefore, reasonable to locate the diaphragm a little lower than the lower focal plane, so that in any case it is outside the lens system. This has the practical advantage that an aerial image of the diaphragm will be projected some distance above the plane of the specimen, where it can be used as a reference point for collimation (Fig. 92).

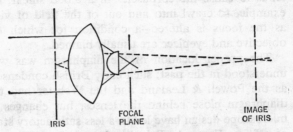

Fig. 92. Aerial image of iris projected by condenser.

It is not, however, reasonable to separate the condenser from the diaphragm by a distance representing a multiple of its focal length, as is commonly done even in some good microscopes. If, for example, the diaphragm is twice the focal length from the equivalent plane of the condenser, instead of once, its image will lie one focal length above the image of the light source (Fig. 93), and the further the diaphragm is displaced, the closer will its image be to that of the light source.

The result is that, except with mathematically parallel light, the substage diaphragm ceases to control only the aperture, and affects the field of view as well, producing a vignetting effect by increasing the

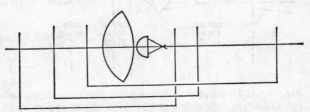

Fig. 93. Image of substage diaphragm approaching condenser focal plane when iris is too far from condenser.

obliquity of the illumination at the margins of the field before the centre is affected (Fig. 94B). This tends to cause the corpuscles in a blood smear, for example, to crawl into and out of the field of view as the focus is altered—a condition for which the objective and eyepiece are usually blamed.

The proper location of the diaphragm was well understood in the past; such early British condensers as the Powell & Lealand and the Webster had the diaphragm close behind the lenses, but changes in microscope design have led to a less satisfactory state of affairs. Formerly British microscope condensers were carried either in a **substage mount** (Fig. 88A),

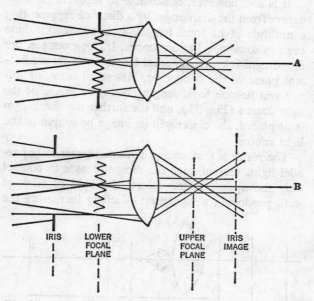

IRIS | LOWER FOCAL PLANE | UPPER FOCAL PLANE | IRIS IMAGE

Fig. 94. A. Diaphragm in lower focal plane producing uniform effect. B. Low diaphragm position affects uniformity of illumination.

which fitted from above downwards into the compound centring substage, with the diaphragm and stop holder immediately below the condenser mounting thread, and their locating sleeve below, or else, in simpler microscopes, in an **understage fitting** (Fig. 88B), designed for insertion from below into a simple tube attached to the stage. In this case the condenser—usually an Abbe Illuminator—was inside the locating sleeve and resting on the diaphragm at the lower end of it. In both cases, therefore, the condenser was properly placed in relation to the diaphragm, and each article worked excellently.

Changes in stand design have resulted in the disappearance of the substage mount, and the universal employment of the simpler type in a compound substage. This is quite satisfactory with the large Abbe and aplanatized types of condensers, which have a long focal length, but when a small fully-corrected condenser has to be used, it is commonly mounted in the same fitting by means of an adaptor. This locates the condenser two or three centimetres above the diaphragm, which is consequently far below its proper position (Fig. 95).

In these circumstances, controlling the aperture without affecting the even illumination of the field is impossible. It is difficult to provide any remedy for this state of affairs, as the available movement of the substage is usually restricted to conform to the manufacturer's own substage mount, and there is seldom room to interpose between the condenser and the standard mount a **Davis shutter,** which consists of an iris diaphragm with R.M.S. threads above and below. If this can be done, however, there is usually a marked improvement. An alternative is to obtain one of the cylinder iris diaphragms formerly made

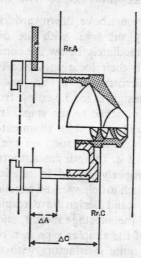

Fig. 95. Relative diaphragm positions for an Abbe Illuminator (A) and a corrected condenser (C) in an understage mount.

by some makers, in which the iris was at the top of an understage fitting, with an R.M.S. thread immediately above it. These may occasionally be found second-hand, but lack any provision for carrying stops.

Chapter Eight

THE ILLUMINATING SYSTEM

Lamps With and Without Optic Axes

The entire illuminating system from the light source to the specimen should logically be considered as a single entity, not as a succession of casual components. The outstanding development of post-war design is the integration of the whole into a convenient and practically fool-proof unit. In order to appreciate the action and control of such a system, however, it is necessary to consider it in terms of its components.

The condenser and diaphragm have already been discussed. The remaining items are the lamp and the microscope mirror.

Microscope lamps can be divided into two classes —those without an optic axis of their own, and those with one. The former are simple luminous sources which radiate light in all directions, and commonly provide a diffuse source of light the area of which can be varied by means of an iris diaphragm. The latter comprise lamps consisting of a small and intense source with a condensing lens to control the divergence of the rays, and a diaphragm to vary the diameter of the condensing lens.

The simpler type of lamp consists in its simplest form of a silica-sprayed or opal electric lamp in a housing which prevents light escaping except through the aperture of the diaphragm, which is close to the lamp surface (Fig. 96). The essential requirement in such a lamp is that the visible luminous area should be uniformly bright. This is the case with silica-sprayed lamps, which exhibit a structureless even

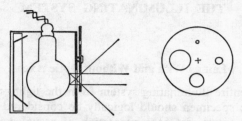

Fig. 96. A simple electric microscope lamp with a field diaphragm.

glow, but frosted glass lamps do not radiate uniformly, and the pattern of the filament is visible. In addition, frosted glass provides a degree of visible structure which makes it necessary to avoid having it in focus with the specimen.

A simple lamp of this type is extremely easy to use, as it essentially presents a luminous disc, which can be reduced to small size and then located on the optic axis of the microscope by movements of the microscope mirror, and subsequently expanded to fill the field of view.

Unless the lamp is so carelessly placed that the diaphragm aperture appears oval instead of round, no further alignment is necessary (see Fig. 99). This type of lamp may be regarded as the modern equivalent of the oil-lamp used by microscopists of the classic era. The features common to both are the structureless and large luminous area, of low intrinsic brilliance, as opposed to the high but localized brilliance of a bare electric filament.

Critical Illumination

The most efficient use of the oil-lamp flame occurs in the conditions called by E. M. Nelson **"critical illumination"**, in which the image of the edge of the flame is focused in the plane of the specimen (Fig. 97). In these circumstances the field of view is un-

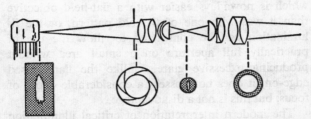

Fig. 97. Critical illumination by Nelson's method, showing light distribution in object and aperture planes.

evenly lit, the lamp flame forming a central stripe, but the aperture of the objective is evenly filled with light, provided always that the substage condenser is a good one. These conditions provide the finest conditions for resolution over a small area, as the objective aperture is completely utilized and under full control. Nelson always considered that only three-quarters of the objective aperture could usefully be illuminated, and that the image deteriorated if this limit were exceeded. He was a most painstaking worker, who based his views on exhaustive trials, but this limitation may be regarded as subject to modification according to the area of the field illuminated.

The critical microscopist was concerned with resolution, and the unevenly lit field was of no consequence to him. The centre of the field was the position for which the objective was computed, and corrections over the rest of it were deliberately sacrificed to ensure perfection at the centre. Anything he examined was, therefore, moved into the centre of the field.

The adjustable circular light source presents practical advantages over the flame in this respect, as it can be made to fill the field of view for searching, which is nowadays easier with a flat-field objective than it was with one of the old pattern, or the illumination can be contracted to allow the use of practically full aperture on a small area without producing excessive glare. Unlike the flame used edge-on, it does not possess a considerable depth of focus, but this is not a disadvantage.

The modern interpretation of critical illumination lays the emphasis on the even illumination of the objective aperture rather than the projection of the radiant image into the field of view. It will generally be found that if even a good condenser is first set to project the image of the lamp diaphragm into the focal plane, full illumination of the objective aperture will demand a slight re-focusing of the condenser. This is due to imperfect correction of the condenser, either as a result of wrong slide thickness or wrong lamp distance, but as long as the optimum position for the objective is secured, the practical result is satisfactory.

Köhler Illumination

When the critical microscopist required to search a slide, and needed complete illumination of the field, he either turned the lamp wick broadside-on, or else used a lamp condenser—a powerful plano-convex

lens called a **bull's eye**—to magnify it. This forms the
basis of the common method of illumination used
today, which is coupled with the name of August
Köhler. It is important to bear in mind that Nelson
was an observer, mainly concerned with resolution,
and Köhler was a photomicrographer, whose first
consideration was, therefore, a uniformly illuminated
field of view, and who required also a brilliant illumination to shorten his exposure times.

In the **Köhler method** (Fig. 98), a well-corrected

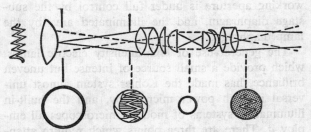

*Fig. 98. Köhler illumination showing light distribution in
object and aperture planes.*

lamp condenser is used to project an image of the
light source into the lower focal plane of the substage condenser. In these circumstances the lamp
condenser will appear from this position as a uniform source of light of great brilliance. All the light
from the lamp condenser enters the aperture of the
substage condenser, so that it is optically economical,
and puts a small condenser on equal terms with a
large one as far as collecting light is concerned. It
does, however, result in uneven lighting of the substage aperture, and therefore of the objective aperture, if the light source itself is not uniform, so that

resolution is not then equal to that in critical illumination.

In those cases where the light source is uniform—where an arc, a Pointolite, or a ribbon-filament lamp is used—the illumination of the substage aperture is uniform, and conditions are thus identical with both Köhler and critical illumination (see Fig. 107).

The substage condenser itself is focused to project an image of the lamp condenser into the field of view, securing uniform illumination over an area equivalent to that of the lamp diaphragm. Thus the working aperture is under full control by the substage diaphragm, and the illuminated area by the lamp diaphragm.

The general use of high-intensity electric lamps, which provide a small source of intense but uneven brilliance, has made the Köhler system almost universal in high power microscopy, and the built-in illuminating systems of modern microscopes all employ it. There are three points which require attention when a separate lamp is used to provide Köhler illumination, and all three refer to the lamp.

Lamp Alignment

A lamp consisting of a small light source and condenser lens has its own optic axis. It is necessary that the lamp filament should lie on the optic axis of the lens (see Fig. 101), and that the optic axes of both

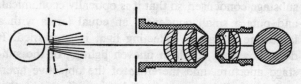

Fig. 99. Luminous source centred to field of view.

lamp and microscope should be properly collimated. This is more difficult than merely centring a luminous source in the field of view, and failure to ensure that it is accurately carried out produces very obvious results, particularly in photography (Fig. 100). Es-

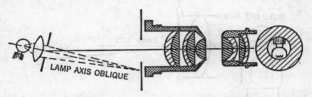

Fig. 100. Projector lamp centred but wrongly aligned.

sentially, the aligning process is simple enough, and consists of centring the lamp diaphragm in the field of view of the microscope, and then ensuring that the lamp filament is also central at the appropriate focal level. The practical consummation of this plan is usually made quite unnecessarily hard by the common habit of mounting lamp-houses by a side trunnion and clamp on a retort stand; if the optic axis of the lamp and its two axes of rotation have no common point, but are separate, collimation must proceed by a series of steps involving the lamp and microscope mirror alternately. When the lamp is so made that it pivots around the centre of the lamp diaphragm, this difficulty does not occur; first the iris is centred, and then the filament brought into the aperture with great exactitude and no delay.

The centring of the lamp filament to the lamp condenser is difficult to carry out with exactitude. Usually the filament itself does not present an ideal object in the optical sense, as it is three-dimensional, uneven, and sometimes not entirely symmetrical.

This is particularly the case when some types of car headlamp bulbs are used. The best method available is to point the lamp squarely at a wall, close the diaphragm, and focus the condenser from one ex-

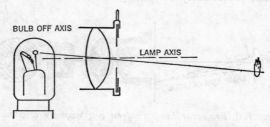

Fig. 101. Centring light source to lamp condenser.

treme of its range to the other. If the filament is on the condenser axis, its image on the wall should not move, but merely alter in size. The lamp holder should be adjusted until this is the case, and care taken that the filament lies squarely across the optic axis, as, if it is obliquely placed, the illumination will be uneven (Fig. 102).

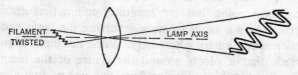

Fig. 102. Effect of a skew filament position.

It is commonly asserted that no great degree of correction is necessary in the lamp condenser. It might be possible to defend this statement if it referred only to chromatic aberration, as so much work requires colour filters, but it is quite erroneous if applied to aplanatism. The whole basis of the

Köhler method is the uniform luminosity of the lamp condenser, and if this is not aplanatic, the entire system is compromised (Fig. 103). The sub-

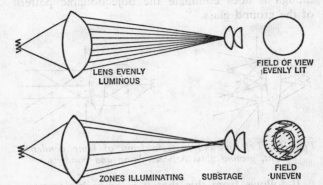

Fig. 103. The effect of spherical aberration in the lamp condenser.

stage condenser will then "see" the uncorrected lamp condenser as a bright central spot and a bright ring, and it will frequently be found that traces of the filament pattern occur in the field of view as well as in the aperture plane.

Use of Diffusers

Most lamps are supplied with a ground glass screen for disguising this condition, but to use one converts the lamp into one of the opal bulb category. The use of a diffusing screen with a lamp condenser is a frank confession of failure, and its use destroys the action of the lamp condenser.

If the ground glass is placed in front of the condenser, it acts as a simple diffuser, just as it would without the condenser; if it is placed between the

condenser and the bulb, the lens becomes the effective source of the diffused light (Fig. 104). The
condenser cannot make the diffused rays parallel,
though it does eliminate the objectionable pattern
of the ground glass.

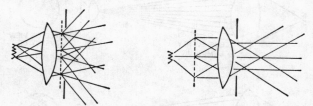

*Fig. 104. Left, ground glass in front of lamp condenser.
Right, ground glass between lamp and condenser.*

It follows from this that it is advisable to use a
lamp condenser corrected for spherical aberration at
any rate, and the distance of the lamp from the substage condenser therefore becomes a matter of consequence, as it should be the major aplanatic conjugate of the lamp condenser. In ordinary conditions,
with a filament lamp, this condition is somewhat
academic, because the projected image of a coiled
filament has a considerable focal thickness. If the
lens is projecting an image ten times the size of the
object—i.e. working on a 1:10 ratio of conjugates—
the axial magnification of the object will be ten
squared, so that the focal depth in the image space is
very considerable.

This has a beneficial effect in reducing the effects
of spherical aberration in the lamp condenser, which
can be observed at their very worst if a pin-hole
diaphragm is placed close to the light source, so
that the condenser is used with a point source of
light.

Aperture and Field Limitations

The real importance of the lamp distance lies in the effect it has on the size of the filament image projected onto the substage condenser. Ideally this should just cover the back lens, so that all the light is utilized. If it fails to do so, the effective aperture of the condenser is restricted to that portion which is covered (Fig. 105A); once the substage diaphragm is sufficiently opened to uncover this area, further opening produces no increase in N.A., as there is no light traversing the outer zones.

It is fair to say that most of the difficulty experienced in arranging Köhler illumination arises from unsuitable lamp condensers. No single condenser fulfills the requirements of providing with optical efficiency the large field of illumination of low N.A.

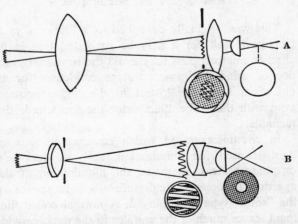

Fig. 105. A. Lamp condenser of long focal length restricting working aperture. B. Lamp condenser of short focal length restricting field of view.

(Fig. 105A) and the small field of high N.A. (Fig. 105B) demanded by low and high power objectives. A compound lamp condenser based on a "zoom" type of construction would be one solution which is bound to be provided sooner or later (compare Fig. 124), but in the meantime the need is met by the lamp designed by Barer & Weinstein and manufactured by the Singer Instrument Company, which has interchangeable condensers (Fig. 106).

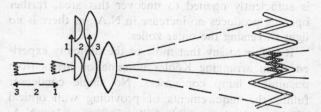

Fig. 106. Lamp condenser of variable focal length to provide both aperture and field of view.

This was originally described as a "rational microscope lamp" but it may be suggested that in its absence microscopists may avoid the irrational practice of using the same substage condenser for all objectives from the lowest to the highest power; even with the Abbe Illuminator the top lens is detachable.

In circumstances in which an annular cone of light is used for illumination, such as dark-field or phase-contrast methods, the illumination of the aperture becomes difficult unless a light source of the "solid" type is employed. A normal coiled filament leaves much of the annulus in the dark, providing bright areas where the filament image crosses the annulus, whereas a ribbon filament or a Pointolite bead illuminates the whole effectively, and provides a considerable increase in brightness and a better image (Fig. 107).

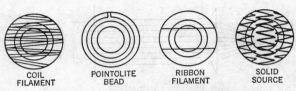

COIL FILAMENT   POINTOLITE BEAD   RIBBON FILAMENT   SOLID SOURCE

Fig. 107. Illumination of a zonal illuminator by various typical light sources.

The Parallel Light Fallacy

Attention may be drawn at this point to the implications of the advice which used to be given in the days of oil-lamps, that when a fully-illuminated field of view was required, the bull's-eye condenser should be used to magnify the light source, which was to be "placed at the principal focus of the bull's eye, so that this projected a beam of parallel light into the substage condenser".

This nonsensical recommendation has been transcribed from book to book, and sometimes even described as "critical illumination with a lamp condenser" to distinguish it from Köhler illumination. Even assuming that the bull's-eye condenser were sufficiently corrected to project a uniform beam of parallel rays from its focal point, it still remains true that for a light source of any finite size, the resultant rays would not be parallel, but would instead be diverging on the paths requisite to produce an image of the source at an infinite distance and of infinite size; the whole concept of parallel rays is an abstraction. If, nevertheless, the substage condenser is lit by

parallel rays, produced by an idealized lamp condenser and a point source of light (Fig. 108), the

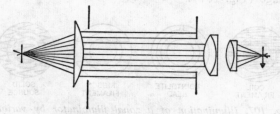

Fig. 108. The "parallel light" fallacy.

image formed in the field of view will be an image
of the original light source, not of the lamp condenser, which will have contributed nothing towards
widening the image. It is not possible to project an
image of the lamp condenser in these circumstances,
except in so far as defects in its surface make it
visible.

In these conditions the action of the iris diaphragm
below the substage condenser and the one in front of
the lamp condenser is identical; both control the
aperture and neither affects the size of the image of
the source, which remains minute and useless for
observation.

The use of "parallel light" is a mistaken concept,
and its recommendation by the classical writers reflects rather the depth of focus of the lamp condenser,
already explained, than their physical appreciation.
A condenser set to focus the near edge of a lamp
flame at infinity—i.e. on the opposite wall—would
focus the remote edge of the flame about twenty
inches from the lamp, so that the effective illumination of the specimen would approximate to Köhler's
method.

It will be appreciated that setting up a microscope properly takes a certain amount of time, and presents certain difficulties and opportunities for error, depending largely on the equipment, and particularly the lamp, available. An intelligent user will find very little trouble, as the requirements are logical and their sequence automatic, but many who use the instrument as a tool are disinclined to adopt a regular practice of preliminary adjustment which appears to them a mere waste of time, or sheer pedantry.

Whilst it may be conceded that for observation somewhat less rigorous adjustment is necessary than for photography, it is a mistake to separate the two sides of the work. If the attitude is adopted which discriminates between visual and projection techniques, the standard of visual work will decline, and it will prove harder to obtain satisfactory results in projection and photography, as the rigid subconscious criticism of conditions automatic in the purist will be absent or unfamiliar just when it would be particularly valuable.

The Microscope Board

There is every advantage in maintaining the microscope and its illuminating system permanently in working adjustment; this saves the time of the expert and preserves the lazy from the effects of their carelessness, whilst ensuring that suitable combinations of equipment are used. Most modern manufacturers supply self-contained microscope outfits, which once connected to a suitable electrical supply provide the optimum conditions for visual work, leaving the observer only to focus the objective and regulate the aperture. On the whole, such instruments are ideal for routine work, where a constant standard of optical efficiency is demanded. The only criticism to which they are open is that they are not suitable for work of a diverse nature; for instance, a really low power with a wide field is not usually available, though medium or high power observation or photography is instantly possible in transmitted light, phase-contrast, dark-field, or incident illumination.

The various complete microscopes of this class cannot usefully be described in general terms, as each represents a complex assembly of optical units specially adapted to it, and the user must make himself familiar with the maker's instructions for the particular instrument. It is worth bearing in mind that these instruments obey the basic requirements of microscopic optics described already, and that a user who knows how the machine functions is in a much stronger position than one who merely knows his instruction book. In spite of the use of lamps in pre-focused mounts it is not impossible to find that the illumination is not properly centred, whatever the book may insist. It is most important that the user should keep his critical faculty unblunted, and on no account allow himself to be blinded by science or pontifical authority, however eminent the manufacturer or the author.

At a lower level of integration than the unit-microscopes there exist the semi-permanent combinations of lamp and microscope represented by the **microscope board** of M. J. Coles, which has recently been revived by Professor G. Needham.[1] This consists simply of a solid wooden baseboard on which

---

[1] Available in a metal form from Bausch & Lomb and American Optical.

the lamp and the microscope are located in definite positions (Fig. 109). Coles used an oil-lamp with a

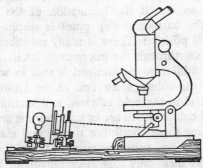

Fig. 109. A simple type of microscope board, with lamp attached and microscope fitting between fixed blocks.

swing-out bull's-eye condenser, and a high screen between it and the microscope to keep extraneous light out of his eyes. The screen formed a filter holder, and served to carry, in a convenient position, a list of magnifications, apertures, and filter factors. His microscope was inclined for use from the vertical position in which a bell-jar covered it between observations. He thus had to align his lighting system to some extent before use, but with modern inclined binocular instruments this is not necessary, and a student's microscope may be left inclined under a plastic cover.

The weak point of this system is the microscope mirror. Most mirrors are easily disturbed, though in some cases locking screws are provided which obviate this, and the collimation of lamp and microscope has to be performed if the microscope is moved.

It is sometimes recommended that a mirror-cover should be used for setting up, the lamp being aimed at a mark on the cover in line with the tube of the microscope. This plan is not a good one; microscope mirrors do not pivot in the plane of their reflecting surfaces, so that if the inclination of the stand is altered, the correct aiming point is displaced from its original position. There is really no substitute for examination through the microscope itself.

The condition to be attained is that in which the image of the substage iris and of the lamp iris are each central in the field of view, and surrounded by fringes of colour which are of uniform hue and width. In obtaining these conditions the ground glass filters supplied for the substage ring and the lamp stop holder find their only logical employment; by using the substage diffuser whilst centring the condenser the effect of oblique illumination is obviated, and by placing it in the lamp, the field diaphragm can be centred with great ease. It should then be removed whilst the lamp axis is aligned, and not replaced subsequently.

It will be found that ordinary transparent light filters are capable of producing an appreciable displacement of the optic axis of the lamp unless they are perpendicular to the optic axis themselves. It is, therefore, best for their carriers to be permanently mounted on the lamp, so that they move with it, and not independently mounted on the baseboard.

Various modern designs of the baseboard, such as the Baker Projectolux base, provide a pre-adjusted Köhler lighting system of variable intensity on a metal baseboard fitted with clamps for the microscope and a rigidly-mounted prism or mirror which provides a vertical beam of light.[2] The microscope

[2] Available in the U.S. from American Optical as an Ortho-illuminator.

is used in a vertical position only, as the necessary inclination is provided by the inclined binocular body, and the illumination is permanently in correct alignment. This method constitutes a versatile and fool-proof combination, capable of visual use, projection, and photography, and is not tied to one microscope only, as one instrument may be removed and another substituted if circumstances demand it. The only point to be checked when this is done, apart from aligning the new microscope, is that the supposed vertical position is actually vertical; in some microscopes the accuracy of the vertical position is not exact, so that unless they are adjusted, the illumination is permanently misaligned.

With any built-in illuminating system, it is worth confirming occasionally that the filament of the lamp remains axial; there is a tendency for the filament to sag as it ages. When a new lamp is inserted it is only prudent to check that its filament occupies the intended position, and that the envelope is free from defects which might distort it.

The Substage Lamp

As an alternative to the use of a baseboard, a small lamp may be mounted in a housing attached either to the substage condenser mount, or in place of the mirror. In the former position control of the illumination is not complete, as the substage diaphragm acts both on the aperture and the field of view, but the excellent lamp produced by Messrs. Beck[3] to replace the mirror provides complete con-

[3] Other excellent substage lamps are available from Zeiss, Leitz, American Optical, and Nikon.

trol of intensity, field of view, aperture, and centration (Fig. 110). It offers the benefits of unit

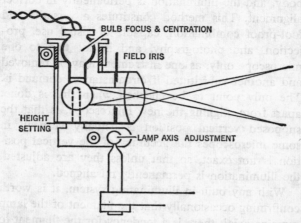

Fig. 110. *Substage lamp permitting complete control of illumination (R. & J. Beck, Ltd.).*

construction cheaply and in compact form; the microscope with lamp attached can be kept in its original case.

Chapter Nine

SPECIAL METHODS
WITH TRANSMITTED LIGHT

With prepared microscopical specimens, the use of a wide, solid, axial cone of illumination provides conditions in which both resolution and visibility are secured. The particular characteristics of the object which assist this effect are those which produce differences of absorption between different parts of the specimen without marked differences of refractive index either in the specimen or at its surfaces.

Specimens are, therefore, normally prepared for microscopical examination by being cut into very thin sections, to make them transparent, stained to produce colour differentiation of their structure, and finally mounted between a slide and a coverslip in a medium which eliminates differences of refractive index between the specimen and the glass.

Circumstances arise, however, in which specimens are ill-adapted to critical microscopy, either from their nature or because preparation is inexpedient. Excluding opaque specimens which require examination in incident light, these are transparent and display marked inhomogeneities of refractive index in their structure; typical examples are living organisms which it is desired to examine alive, and aggregates of material in which the texture is spongy and the surface irregular.

Critical images of such specimens are not very informative, as the image contrast is low and glare is usually present. It is, however, possible to produce useful images in many such cases without departing from the conventional condition of image-formation, namely that the reconstitution of the rays into the image should preserve their original relationships.

If the wide solid axial illuminating cone is unsuitable, the possibilities of narrow, oblique, and hollow cones must be explored, and combinations of both groups are often useful.

The use of a narrow axial cone of light has been mentioned already; it produces increased visibility at the expense of definition. The transparent fibre which is invisible in critical conditions shows as a dark band, accompanied by diffraction fringes, and isolated particles appear swollen, with haloes around them. Such images are to be accepted as representing the fibre or the particle in its position, but in nothing more; width cannot be directly determined.

Ambiguity sometimes exists between artefacts such as the diffraction bands accompanying a fibre in these conditions, and a true sheath or capsule. The question of the identity of a doubtful case can often be resolved by using coloured light. A capsule is a definite entity of a definite size, whilst a diffraction halo represents a reaction to light of a particular wavelength, and changes its position if the wavelength is changed, moving further away in red light and closer in in blue light. A parti-coloured lamp filter can be used to examine the behaviour of questionable examples; anything which traverses the junction from red to blue-green without deflection has real structure.

Narrow-cone illumination has no real value as a method of examination; it provides an opportunity to see inconspicuous features, but should be abandoned once the feature is detected, as it can provide nothing further except misleading impressions. A feature once noticed can usually be seen when the diaphragm is opened to a reasonable extent, and living organisms are usually visible as long as they are moving.

Olique, Annular, and Dark-Field

Oblique lighting (Fig. 111A) may be rapidly dismissed. It is the oldest and simplest method of providing some indication of solidity in a specimen, and may be used at low powers to demonstrate the shapes of cavities or solids, owing to the uneven lighting of their sides, but has been almost entirely superseded by the excellent stereoscopic effects produced by low power binocular microscopes. It may still be regarded as an expedient suitable for certain conditions in photography.

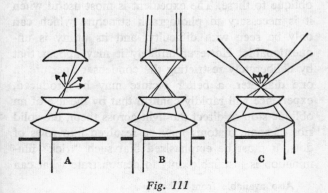

Fig. 111

Formerly it was produced by moving the microscope mirror bodily out of the optic axis, but present-day design employs a fixed position for the mirror, and oblique light is obtained by means of a sectorial stop in conjunction with the substage condenser. Some European Abbe substages make provision for oblique illumination by mounting the iris diaphragm on a plate which can be displaced sideways and rotated about the optic axis.[1] The only disadvantage in this convenient arrangement is the tendency for the position of axiality to become somewhat vague with use, though the modern pattern should avoid this.

Oblique lighting is employed at high powers to improve the visibility of certain parts of the pattern of markings of diatoms, which are regularly disposed in closely adjacent lines. Essentially, this involves producing a radial variation in the resolving power by using a stop either sectorial or radially slotted in the azimuth appropriate to the specimen. The effect is to increase the resolution and enhance the visibility of features which lie in one direction by diminishing that of others which are transverse or oblique to these. The expedient is most useful when it is necessary to photograph structure which can only be seen with difficulty, and its utility is unquestionable. However unlikely it may appear that by deliberately restricting the condenser aperture in one diameter, a better picture may be produced, experience will rapidly confirm that by the use of an oblique stop to direct the light across them, the solid ribs of some diatoms may be resolved into rows of dots. It must be emphasized that such "trick" illumination is justifiable only to demonstrate what can

[1] Also available from Nikon.

be proved by other means; it cannot be relied upon as a means of research.

The other special methods available with conventional apparatus are **annular illumination** (Fig. 111B) and **dark-field illumination** (Fig. 111C). The former may be considered as oblique lighting made radially symmetrical, so that the image is not influenced by the orientation of the structure of the object. The illuminating beam is a hollow cone, produced by a substage condenser having the central portion of its aperture obscured. The size of the portion obstructed is a matter for experiment in particular circumstances, whilst the aperture of the total illuminating cone may be equal to, but should not exceed, that of the objective in use. The method is applied as a rule to the study of diatoms, and obtains its advantage over a solid cone from two causes.

A bright point in the object is imaged as a bright disc surrounded by rings of rapidly diminishing brilliance if the aperture is unobstructed or the illumination is by a solid cone of light. If the aperture or the illumination is annular, the diameter of the disc is reduced, whilst the rings are brighter. In visual use there is little to be gained from an annular cone of light, but photographically a condition is possible in which with a short exposure, only the central disc registers on the plate, and abnormal resolution can be achieved.

The second effect which may be produced depends on the distribution of spherical aberration in the objective. If zonal errors exist, an annular illuminating system may produce increased visibility by introducing phase contrast in the image; the basis of this method will be explained later, but it may be loosely described as a deliberate zonal spherical error in the objective which causes an invisible pattern of phase differences in the object to be imaged as a visible pattern of intensity differences. It was believed formerly that an objective especially good for diatom examination was by no means the best for ordinary high power microscopy, and vice versa. The use of annular illumination was then very popular, but the difference between visibility and resolution was generally ignored, so that circumstances which promoted visibility were commonly considered to increase resolution, and it was quite general for an objective to be described as useless with a solid cone of light, but marvellous in annular illumination for diatoms, though not for ordinary objects. The essential fact is that diatoms, being very regular in their structure, produce distinct and localized diffracted rays, which use only definite parts of the aperture, whereas non-periodic objects tend to fill the entire lens with light diffracted from their irregular features. Consequently a certain degree of local error might be innocuous or even helpful with certain diatoms, though with general objects or diatoms with differently spaced patterns, the image might well be useless.

Annular illumination, like the asymmetrical oblique variety, must be regarded as a method of enhancing visibility in suitable circumstances, and not as a reliable method of investigating the unknown.

A type of substage condenser is produced by several firms which provides an annular cone of light of variable obliquity. Originally these were intended for use with phase-contrast objectives of various focal lengths, but their use is sometimes recommended to secure pseudo-stereoscopic, or "plastic" effects with ordinary objectives. The idea is by no means new, as a variable-sized central stop like a reversed iris diaphragm was available in the past for this purpose, and the fact that it is no longer manufactured indicates that the results obtained with annular lighting were not sufficiently useful to be irreplaceable. Like so many things in microscopic illumination, annular illumination has its convenient uses, and may be employed provided the user realizes what he is doing, and refrains from considering that the image which it produces at his hands is necessarily characteristic of the object in other circumstances.

Dark-field Illumination is in a different category of utility. It is a reliable technique of long standing, which enables a sharp image to be produced by tenuous structures which would in normal observation remain invisible. Essentially it depends on image formation by diffracted rays in the absence of the direct beam, and provides high resolution combined with high visibility in circumstances which are unlikely to be mistaken for a direct picture of the object.

The basic technique is simply to use annular illumination at such an obliquity that the direct light does not enter the objective, but only light which has been diverted by the specimen. This produces a luminous image on a black background of those parts of the specimen which show abrupt changes in refractive index—fine threads, flagella, folds in membranes, the edges of transparent sheets, etc.—but whose images would be swamped in the general illumination in bright-field conditions.

As these images are bright on a dark field, whereas in ordinary transmitted illumination they would be dark on a bright field, it is clear that the combination of dark-field and ordinary illumination, as for example by the use of a condenser aperture larger than that of the objective, will produce a degraded image. This is a circumstance to be avoided with variable annular illuminators.

The use of dark-field illumination, particularly at high powers, shows up errors of correction in objectives very plainly. For this reason high power work is best carried out with fluorite or apochromatic objectives in homogeneous immersion, to eliminate as far as possible both chromatic and spherical aberrations.

Low power dark-field illumination is produced very simply by obstructing with a central stop the aperture of the substage condenser corresponding to that of the objective in use. The illumination is set up and centred as for normal use, the substage diaphragm opened wide, and a stop inserted in the substage ring. These stops are commonly either circular black discs on a glass disc, or else metal discs with radial locating arms and a rim to fit the stop tray, and they can be made or obtained in various sizes. As convenient a plan as any is to cut discs from black paper with a set of cork-borers, and to attach them with an adhesive gum to glass or lucite discs of the appropriate size, taking care that they are centrally placed.

If too small a disc is used, a rim of light will be seen at the margin of the objective if it is inspected from above, and instead of being dark, the field of view will be merely dimmed. A stop should be chosen large enough to make the field quite dark, but not excessively large, or the residual condenser aperture available for illumination will be restricted. With an Abbe Illuminator, it will be necessary to raise it somewhat above its usual focal position to bring the marginal focus into the plane of the specimen, and indeed even with an aplanatic condenser some adjustment is commonly required.

Dark-field illumination is useful where considerable differences of refractive index exist, as in specimens mounted in water, or else dry; specimens in resinous media are not well shown by this method, as the discrepancy in refractive index is slight. It is traditionally applied to the study of pond-life, which provides images of great beauty as well as great clarity by this means.

The method described of obtaining dark-field illumination with a substage condenser and central stop naturally becomes less effective as the objective aperture increases and the margin remaining for illumination decreases. With the Abbe Illuminator, it is possible to use objectives of 0.65 N.A., particularly if the condenser is immersed to increase the light transmission. If a properly corrected immersion condenser is employed, it is not impossible to obtain a dark-field at 1.0 N.A., though it is very inconvenient and need not be attempted if the stop holder is not immediately behind the condenser lenses.

Reflecting Dark-Field Illuminators

For use with these higher powers, reflecting condensers are always used in practice. These invariably work in oil immersion, and provide annular illumination between the limits of about 1.1 and 1.5 N.A. according to their design. Most are designed for use with an objective of 1.0 N.A., which may be a special fluorite objective of 50x designed for this use, or a 90x with a built-in iris diaphragm or other means of restricting its aperture.[2] A very few dark-field illuminators will provide a dark field with the full aperture of a 1.4 N.A. objective, thus eliminating the loss of resolution caused by reducing the aperture, but these are not commonly used, as they require a mounting medium with a refractive index exceeding 1.4 to transmit their light, and the ordinary field of employment of high power dark-field illuminators is the examination of aqueous mounts in pathology, where the N.A. of the transmitted light is restricted to 1.33 by the refractive index of the water.

The oldest type of dark-field illuminator is the **paraboloid.** It is a property of reflection at parabolic mirrors that light parallel to the axis is accurately reflected through the focal point, even when the mirror is made with a considerable aperture. For microscopic purposes the mirror takes the form of a solid glass block with a parabolic surface which reflects the light by total internal reflection (Fig. 112A). The curved surface is truncated so that the focus lies in the plane of the specimen when a slide is placed in immersion contact with the paraboloid. Unfortunately, as their aspheric surfaces cannot be ground to the same degree of precision, paraboloids are less accurately made than other dark-field illuminators. Nevertheless, they provide excellent dark-field effects and are easy to use, as they will work through relatively thick slides and are not very difficult to centre. They also enjoy the property that if used over an iris diaphragm, their minimum working aperture can be increased by closing the diaphragm, which cuts off the less oblique marginal rays before the more oblique inner ones.

[2] Funnel stop.

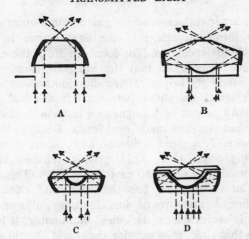

Fig. 112. High-power dark-field illuminators.
A. Immersion paraboloid, with iris (Bausch & Lomb).
B. Zeiss Cardioid.
C. Leitz Concentric.
D. Cooke, Troughton & Simms adjustable pattern.

Modern thought favours doubly-reflecting illuminators, (Fig. 112B–D), as these can be given a
considerable degree of spherical correction, and
therefore provide a greater concentration of the light
at higher apertures than the paraboloid. It is possible
to obtain a recognizable image when some of them
are used as objectives. The simplest type of this pattern is the concentric illuminator introduced by Leitz
many years ago (Fig. 112C), and modified by most
manufacturers since. This consists essentially of a
glass plate with a polished hemispherical depression
in the upper surface, and a polished hemispherical
outer surface concentric with it, except below the
inner depression. The top is covered by a glass plate.
Light entering the illuminator from below is reflected

by the inner surface onto the surrounding outer one, which forms an image of the light source in the plane of the specimen. The focal length of the combination is short, so that the image is small, and correspondingly bright. These illuminators are usually designed for slides between 0.5 and 1.0 mm thick, and cannot be brought to a focus with thicker slides, but patterns made by Messrs. Beck and also by Cooke, Troughton & Simms, Ltd., incorporate an adjusting collar (Fig. 112D), which enables them to be used with slides up to 1.5 mm thick. This is a great advantage, and has the additional effect of providing a measure of fine focusing adjustment, which is useful with a dark-field illuminator. If used with a thin slide, they provide dark-field illumination with objectives up to 1.2 N.A.

The essential requirements for using high power dark-field illumination in the case of conventional types of microscopes, in which the illumination is not integrally included in the design, are a powerful light source, preferably a Pointolite or ribbon-filament lamp with a lamp condenser, and the ability to centre the illuminator to the objective. The principal difficulty lies in centring the illumination, as errors in this respect produce a chaotic picture, which cannot be used.

Centration

The difficulty in centration arises from the lack of a direct axial beam of light; the components have thus to be aligned by indirect means. If a specimen is already being examined by direct light, with the optical system properly centred, it is not difficult to centre a dark-field illuminator if this can be substituted for the ordinary substage condenser without disturbing the microscope mirror, but if this is impossible, the operation must be commenced from the beginning.

When dark-field illumination is required from the outset, without using direct light, a medium power dry objective should be used for setting up. The microscope should have its stage horizontal, the dark-field illuminator should be inserted, and its upper surface examined to locate the ring which is commonly engraved on it to indicate its centre. This is brought into the field of view and centred by using the substage centring adjustment. The illuminator should be lowered a millimetre or two below the surface of the stage, and a drop of oil then placed on its upper surface. Another drop is placed on the lower side of the slide, which is carefully placed on the stage, and contact made between both drops by raising the illuminator. This ensures that air bubbles are not trapped in the oil layer, where their presence reflects light and prevents a dark field being obtained.

The lamp is focused to project an image of the light source onto the microscope mirror, and this is adjusted until a reflected image of the lamp iris is located on the front of the lamp itself (Fig. 113). This image arises from back-reflection from the upper surface of the slide, which is illuminated at more

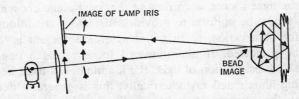

Fig. 113. Collimation of lamp axis by back-projection.

than the critical angle; its size can be varied by slight alterations of the substage focus as well as by adjusting the lamp iris. The reflected image is manoeuvred by adjustments of the mirror and the lamp alignment until it is exactly coincident with the lamp diaphragm; this represents an exactly axial alignment of the lamp and illuminator, whilst the illuminator has itself been made coaxial with the objective by the substage centration. The adjusting process is made much easier if the lamp is constructed to pivot about the centre of its iris than it otherwise may be.

The slide is now examined with the medium power objective, and will show a ring of light which contracts to a bright spot as the substage focus is altered; if it cannot be made to contract to a uniform spot, the slide is too thick, and cannot be used.

The high power objective is now substituted for the medium power, and the bright spot located and centred by means of the substage centring screws. If it is not visible in the field of view initially, the position of the spot may be found by focusing the illuminator down to produce a ring; a sector of this will come into view, and the illuminator may be centred and refocused from the guidance this provides.

The bright spot is a good image of the lamp iris or the light source, according to the focal plane chosen. It must be uniform to provide satisfactory conditions for observation, so that a structureless radiant is to be preferred. If alterations of focus reveal a one-sided distribution of light, the illuminator is itself being illuminated asymmetrically; this is generally due to the lamp filament being off-axis with respect to the lamp condenser-mirror-illuminator line, and it should be adjusted accordingly.

The complex process of alignment has always been an obstacle to the use of high power dark-field illumination, and was one of the principal reasons for the replacement of the microscope mirror by a small lamp attached directly to the substage, which eliminates most of the difficulty. In modern unit-constructed microscopes, particularly the Zeiss stand, in which the conventional substage focusing system is abandoned, the process is much easier; the condensers are exchanged by an interrupted screw thread, and the dark-field illuminator focused immediately by a radial lever, the centration being automatic and extremely exact.

Owing to the high visibility obtained against the dark background, extraneous material becomes very evident in these conditions. Imperfections in the slides or coverslips cause severe glare, and reflections from the objects in the field of view may become embarrassing by their effect on each other. Specimens should, therefore, be mounted between very clean and perfect glasses, and kept as thin and sparse as possible.

The Axial Dark-Field System

There is a variation of dark-field illumination in which the illumination is axial, and the centre of the objective is obstructed, the image being formed by the diffracted rays which pass around it. The condition is, in fact, that which is obtained by using a dark-field illuminator as an objective and an objective as the illuminator. It was proposed by J. W. Gordon in the course of a somewhat bitter dispute about the Abbe theory of vision, and condemned as providing conditions in which the first diffracted band of rays might easily be excluded from the image, producing a spurious doubling of the actual number of lines present in a regular pattern. It was used by Rheinberg in 1896 in slightly different conditions, and has more recently been revived in connection with reflecting objectives similar in design to dark-field illuminators. It is generally conceded that the results obtained differ somewhat from those obtained by annular dark-field illumination, and it has been suggested that this is because diffracted beams are either excluded entirely or else admitted to the objective as a complete set, whereas in the case of annular dark-field only part of the diffracted beam of any given order is admitted. If this is correct, the image might be expected to be either false, or else more accurate than the common one. It is, therefore, a process to be used with caution; it may improve visibility, but it is necessary to check that what it reveals has actual existence.

This is also true of the Spierer system, which combines a peripheral and also an axial dark-field system simultaneously with incident light reflected down onto the specimen from a silvered area in the centre of the objective lens. It is claimed that this shows structure which no other system reveals, but it is asserted by hostile critics that the appearances produced are at least partly independent of the specimen, and are due to interference effects in the illuminating beam itself.

Rheinberg's Differential Illumination

A modification of the dark-field system was produced by Julius Rheinberg in 1896, and is commonly named after him. In this, which resembles low power dark-field illumination, the central opaque stop used under the condenser is replaced by a transparent coloured filter (Fig. 114), which transmits a coloured background against which the specimen is seen illuminated in white light or in the colour of the

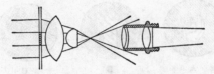

Fig. 114. Arrangement for obtaining colour contrast by Rheinberg method.

peripheral portion of the substage stop. Generally this is complementary in colour to the background, and sets of Rheinberg discs can be made with red/green, blue/yellow, or other combinations to suit particular objects. As the central illumination is the more intense, the central stop must be darker than the peripheral zone if the effect is to be marked.

It may be noticed that although in these conditions the image is formed by rays both within the N.A. of the objective and also outside it, the fact that they are of different colours prevents the degradation of the image which results from a mixed direct and dark-field image. Two superimposed images are formed by different means, a dim absorption image in direct light of the colour of the centre, and a brilliant dark-field image in the colour of the margin, which overwhelms the image due to the direct light (Fig. 115B). There is no mutual interaction between the two images unless the direct light is powerful enough to colour the image appreciably, which should not be the case; it is intended to provide no more than a background glow.

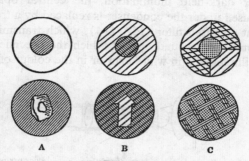

Fig. 115

A. *Rheinberg stop with Polaroid centre and appearance of image with reduced intensity of direct illumination.*
B. *Bi-coloured Rheinberg stop and image appearance.*
C. *Quadrantal stop and appearance of interlaced fibres.*

In the case of commercial sets of Rheinberg discs the centre is often too brightly transparent. Rheinberg himself insisted that the centre disc should never be ground, to act as a diffuser and reduce its brilliance, though the objection would be debatable.

If a blank glass disc is fitted with a central stop of Polaroid, any condition can be obtained from true dark-field to normal illumination by means of a second Polaroid disc on or in the eyepiece. This provides conditions for a coherent background to be obtained, as the direct and indirect lighting is capable of interference, and the resulting brilliance represents the square of the added amplitudes. A very slight departure from absolute dark-field conditions often makes the image more intelligible in this case, though the visibility is somewhat less extreme than in a true dark-field (Fig. 115A). Such a polarizing disc permits the relative intensities of the "Rheinberg image" to be adjusted with a normal set of stops.

The advent of phase-contrast techniques has rendered the Rheinberg system obsolete for serious work, but it is still of interest as representing a very simple method of producing colour contrast in transparent specimens with conventional objectives, and it provides a convenient method of demonstrating the presence of fibrous structures in perpendicular directions. By using a disc in which the periphery is divided into quadrants in alternate colours, these inclined fibres can be made to appear in the different colours of the quadrants opposed to them (Fig. 115C).

This completes the description of image-formation by conventional methods in transmitted light, and it is convenient at this point to summarize the characteristics of the various conditions encountered.

Objects exhibiting variations in absorption but not in refractive index are best suited to direct illumination by solid cones of light, whilst transparent objects with abrupt changes of refractive index are well-displayed by indirect methods such as dark-field or Rheinberg illumination.

In either case, the image is formed in conformity with the rule that the rays participating in it should be reunited in their original phase-relationships. This concept may be called the **Conventional Condition**, and is intended to ensure that the image should represent the object characteristics as faithfully as possible. Deliberate violation of it must be expected to yield a different result, and requires special consideration.

Image Formation by Phase Contrast

The principal reason for the careful correction of objectives is to ensure that the phase-relationships of the image-forming rays shall be accurately maintained, and the image reliable. It is, however, possible to deal with the difficult situation in which the object is characterized only by small variations in refractive index by producing an image which differs from the object.

The phase-contrast method depends on a difference in behaviour between transparent and opaque objects when they diffract light. An object which is transparent and opaque in different parts can be seen by the process of absorption; the opaque parts obstruct the passage of light, and the image consists of a pattern varying in brightness, which is readily visible (Fig. 116). It is therefore called an **amplitude object,** as it affects the amplitude of the light rays traversing it.

On the other hand, an object consisting of transparent features which differ only in refractive index does not affect the amplitudes of the light rays, though it causes them to differ in phase according to the path they have taken through the object. The image conventionally formed by such rays consists of a pattern of phase differences of uniform brightness, and is invisible to the eye or the camera (Fig. 117). Such an object is termed a **phase object,** and the majority of living cells conform to this category. In some cases polarized light or the use of dark-field illumination provide a visible image, but in

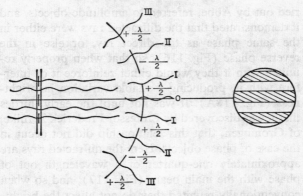

Fig. 116. An amplitude object, showing the phase relationships of the main and diffracted rays and the appearance of the image.

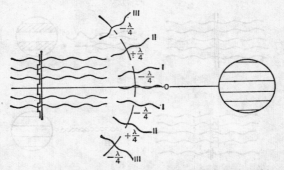

Fig. 117. A phase object, showing the phase relationships of the main and diffracted rays and the appearance of the image.

the majority of cases, the only worth while process formerly available was to convert them into amplitude objects by staining them, which involved killing them and producing artifacts in their structure in the process.

The classic work on the diffraction theory, carried out by Abbe, referred to amplitude objects, and it demonstrated that the diffracted rays were either in the same phase as the direct ray, or else in the reverse phase (Fig. 116), so that when properly reunited with it they would either reinforce it or interfere with it, producing a visible variation in brightness (Fig. 118A). It was not until the early thirties that it was discovered by Professor Frederick Zernike, of Groningen, that this relationship did not occur in the case of phase objects. Here the diffracted rays are approximately one-quarter of a wavelength out of phase with the main beam (Fig. 117), and so when conventionally reunited they do not affect the brightness, but only the phase relationships in the image (Fig. 118B).

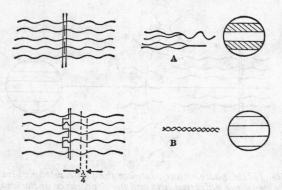

Fig. 118. A. Amplitude object with image formed by interference. B. Phase object with image showing variations in phase, not intensity.

The time-honoured method of forming a dark-field image by using only the diffracted rays, which can interfere mutually, is very sensitive, but is subject to certain limitations, particularly in the features of the object which can be made visible. It shows localities in the object where there is an abrupt change of refractive index, and so produces similar images of a fine fibre, the edge of a sheet, or a fold in the sheet. It is excellent for showing bacteria, but shows also ultra-microscopic particles, so that a turbid specimen may be masked by a haze of light.

Zernike produced a visible image in these circumstances by deliberately advancing or retarding the main beam, after it had traversed the specimen, by one-quarter of a wavelength, without disturbing the diffracted rays. Consequently, when the whole beam was reunited, conditions for interference existed, and the transparent specimen produced an image of the type which it would have produced if it had been an amplitude object, and thus showed differences of optical path length through the specimen—refractive index differences—as though they were differences in transparency.

The necessary adjustment of the phase relationships is carried out by very thin films of material deposited by a process similar to lens-blooming. It is possible to produce a phase delay without absorption, or an absorbing layer with negligible phase delay, by using cryolite or metal for the blooming process; layers of each type can be made either contiguous or superimposed according to the need (Fig. 119).

The differential effect between the direct and the diffracted rays is obtained by locating the film at the upper focal plane of the objective, where the direct and the diffracted rays traverse different parts of the aperture. If an object were to be illuminated by parallel axial rays, they would be brought to a focus at the upper focal plane of the objective, the direct

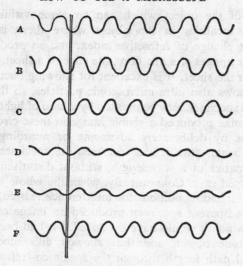

Fig. 119. Effect of coatings producing
B. Retardation only.
D. Absorption only.
E. Retardation and absorption.
A, C and F. Uncoated zones.

rays in the centre of the lens and the diffracted rays
outside the central region (Fig. 120). Thus, by depositing a central disc, or omitting a central disc
from a complete coating, a path difference could be
introduced between the two sets of rays.

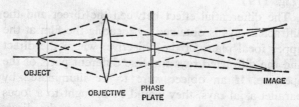

Fig. 120. Phase control by localized retardation of image-
forming rays.

To improve the resolving power, which is poor with axial illumination, a ring-shaped phase-changing zone is used in practice, and the objective is illuminated by a hollow cone of light provided by a condenser fitted with an annular stop which is centred to the phase ring in the objective (Fig. 121). The phase ring is located at a level conjugate with the condenser stop, so that when properly set up the two can be accurately superimposed.

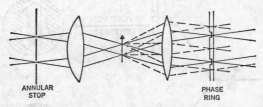

ANNULAR
STOP

PHASE
RING

Fig. 121. Practical application of method of phase adjustment.

In these circumstances, the direct light all passes through the phase ring, and practically all of the diffracted light misses it. The annulus is usually made about one-third the diameter of the objective lens, and with a width of about 0.05 N.A., though these values are varied in certain circumstances.

Usually the phase-changing layer is deposited on the face of one of the components of the objective where it can be sealed between two lenses and preserved from alteration.

The direct beam is always much brighter than the diffracted rays, and the visibility of the image is improved if it is partly absorbed, to produce a better balance between the intensities of the interfering components. This is effected by depositing a partly opaque annulus, the retardation of which may be ignored. If the direct ray is to be retarded also, the retarding layer is also deposited on the annulus, but if it is to be advanced, the retarding layer is applied to the remainder of the objective surface instead.

The difference in the arrangement is shown in the image; and advanced main beam shows high-refractive index areas of the object as darker areas in the image, whilst a retarded main beam shows them brighter than the field (Fig. 122).

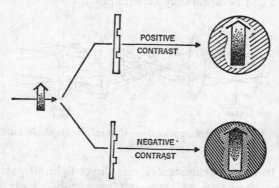

Fig. 122. Appearance of an object of uneven thickness in phase contrast.

It follows that the exact significance of the distribution of density in the image depends largely on the specification of the phase plate, and that the same specimen can be made to show differing appearances with different objectives. It is customary to describe phase-contrast objectives as **positive** if they show high refractive indices darker than the field, and **negative** if they show them brighter; the former is generally preferable, but the latter condition may be more immediately apparent, and so more convenient for searching purposes. The image is also affected by the relative brightness of the two components, and this is specified by the **transmission ratio** of the absorbing layer.

The primary use of the system occurs in the case of small phase differences in the object, and where large values are encountered, exceeding half a wavelength, the contrast may be reversed in the image; a feature three-quarters of a wavelength retarded is equivalent to one a quarter wavelength advanced. Some objectives for special purposes are made with the absorbing layer applied to the diffracted rays for use in such conditions, but they are of very limited application, and that usually where conventional methods would be suitable. In fact, where phase contrast runs into difficulties owing to excessive differences of refractive index, normal image-forming phenomena are quite satisfactory, and these can be employed by simply removing the annulus from the condenser. Special systems have been produced which provide a variable retardation, to enable individual features to be shown to the maximum extent, but it has been found that these are rarely as helpful as would at first sight appear, and the results require very careful appreciation; this type of application is better suited to the **interference microscope.** Phase contrast is ideally applied to the study of living cells, where it can be used qualitatively, and measurements made in less ambiguous conditions.

It is worth noticing that whereas any objective is limited in its resolving power by the necessity of including at least two components of the diffraction pattern in its aperture (Fig. 123A), the phase-contrast objective has an *upper* limit to the size of structure which it will show *in phase contrast,* as unless the diffracted rays are sufficiently deflected to miss the annulus (Fig. 123B), they will not be altered in

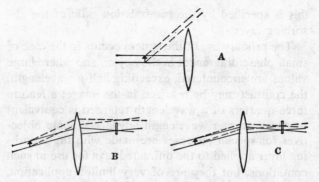

Fig. 123. The limits of resolution in phase contrast.
A. Total failure of resolution owing to inadequate N.A.
B. Resolution in phase contrast.
C. Resolution but no phase contrast.

phase. The narrower the width of the annulus, therefore, the larger the features which will be affected. Anomalous results could occur where the diffracted rays pass through the opposite side of the annulus, and it would appear on this basis that a larger annulus diameter than that commonly used would be preferable.

The practical use of phase contrast presents little difficulty. Most equipment comprises a set of objectives with their phase-plates in position, and a sub-stage condenser of restricted aperture with a series of annular stops. The objectives are mounted and set up for critical illumination without any obstruction in the condenser aperture; it is desirable that the image of the light source in the objective should cover the whole lens evenly, and not consist of a criss-cross of incandescent filaments (Fig. 107), which provide uneven illumination of the annulus.

The annular stop appropriate to the objective is then placed under the condenser, where it occupies a defined position and projects an image which falls onto the objective annulus. The coincidence of these two is important, and is checked by the use of a sighting telescope instead of the normal eyepiece, or by means of a Bertrand lens if one is fitted; lack of coincidence is corrected by the screws centring the annulus, and difference of size between the two by adjustment of condenser focus. It is best to carry out the centration and focusing through the slide to be used, with the object outside the field of view.

Exchange of objectives demands exchange of stops and the checking of centration in most cases, which is a nuisance, though some manufacturers avoid this; Beck uses stops in individual trays, which are centred permanently. The Cooke design embodies a condenser which is not centrable and an iris and a disc of stops which are both independently centrable.[3] When a centring objective changer is used with this assembly it presents severe centring problems initially, but thereafter they do not arise; the essential thing is to centre the substage diaphragm to the condenser to start with, by a back-projection method (cf. Fig. 113), and use this to centre the objectives.

The Zeiss Jena pancratic substage provides an elegant method of dealing with the change of substage stops (Fig. 124). In this, a variable magnification system projects an image of the iris diaphragm and stop tray into the back focal plane of the condenser. Ordinary adjustments of substage aperture are made by varying the size of the projected image of the iris, and the image of a single annular stop placed on the iris can be expanded or contracted to suit the re-

[3] Also Zeiss, American Optical, Bausch & Lomb, and Nikon.

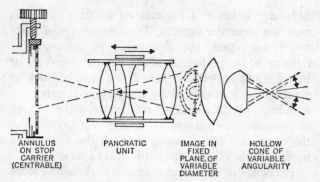

Fig. 124. Use of a pancratic unit to illuminate any desired zone of condenser (Zeiss).

ANNULUS ON STOP CARRIER (CENTRABLE)   PANCRATIC UNIT   IMAGE IN FIXED PLANE, OF VARIABLE DIAMETER   HOLLOW CONE OF VARIABLE ANGULARITY

quirements of the objective. This is also notable as producing an illumination by a definite zone of the condenser, which alone is illuminated, and not by the entire condenser projecting an image of a frosted annulus in the substage.

A somewhat similar device is produced by the Officine Galileo.

Interference Microscopy

Basically, all microscopic image formation depends on interference; in conventional circumstances this occurs automatically between the components of the illuminating beam which are diffracted by the object, and in phase contrast the same result is produced artificially. The term Interference Microscopy is by convention restricted to the situation where light which has traversed the specimen is made to interfere with light which has not done so (Fig. 125). This provides a means by which the thickness

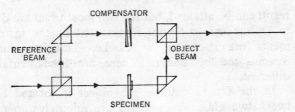

Fig. 125. Optical principle of interference microscope.

and refractive index of the object can be measured; basically this involves compensating one of a pair of exactly equal light-paths for an addition to the other represented by the passage through the specimen.

The illuminating beam of the interference microscope is, therefore, duplicated, and two images of the light source superimposed. These images are adjusted ideally to differ by half a wavelength, and thus to cancel out. If an object is introduced into one illuminating beam, the adjustment will be upset, and the object will appear bright on a dark field. By adjusting the compensation in the other beam, any part of the object can be brought to a state of invisibility, and the degree of compensation causing this will represent the optical thickness of the particular part of the specimen.

It is thus possible to make measurements well within the range of half a wavelength in thickness. By employing a gradually increasing compensation from one side of the field of view to the other, so that a series of parallel interference bands crosses it, and is displaced by the object, a very much finer measurement still is possible, by means analogous to the estimation of heights on a contour map (Fig. 126).

There are several optical methods by which this

result can be attained, but the two most usual are due to Dyson and to Smith, represented by the instruments manufactured by Cooke, Troughton and Simms, and by Baker.[4] These are fundamentally different.

In the Dyson pattern, the object is located between two glass plates which are minutely tapered in thickness, and semi-silvered on all four faces. Light from the substage condenser is partly reflected at each surface, so that the light source is focused both in the specimen and also below it, and the light from both images is re-imaged at a higher level, where the objective is focused, as in the Dyson long-working-distance attachment. This provides conditions in which measurements of optical thickness can

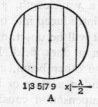

$1|3\ 5|7\ 9 \quad x|\frac{\lambda}{2}$
A   B

Fig. 126
A. Empty field of view showing interference fringe.
B. Effect of introducing a specimen.

be made by traversing one of the tapered plates by means of a micrometer screw.

The Baker-Smith instrument essentially depends on the use of polarized light to provide a bi-focal condenser and bi-focal objective. Below the condenser is a crystal plate, which splits the light into two perpendicularly-polarized components with different foci, which are displaced either axially or else later-

[4] Available in the U.S. as American Optical Baker.

ally, according to the construction chosen. Above the objective their relative phase-differences can be varied by a crystalline compensator, which presents a different path length to the two polarized beams, and these are then re-combined by a further crystal plate corresponding to the one below the condenser.

It is important to realize that the contrast in the image is not due to double refraction in the object; the system exploits the possibilities of polarized light to control the image, but not to form it. Apart from its possibilities in making accurate measurements of optical thickness, the instrument can be used to obtain colour contrast between the object, or parts of it, and the field, and changes in the object are represented by changes in its colour; isolated cells, yellow against a blue background, will individually become red or green and abruptly vanish as they die and disintegrate.

An important advantage of interference microscopy over phase contrast lies in the elimination of the "edge-effect". This is a halo which surrounds isolated objects seen by phase contrast, and represents an excessive reaction from the margin of the object. The image in phase contrast is due to light which passes round the object, light diffracted from its margins, and light diffracted from its interior structure, and these interact to produce an image in which the margin is represented by a bright zone passing into a dark one, and both reverting to an even hue inwards and outwards. This masks the true extent of small objects, and leads to the opinion that a phase-contrast microscope is incapable of producing a clearly focused image.

The interference microscope avoids this appearance, and maintains the effect due to the thickness of an object over its entire extent, not merely near its margins. Part of this advantage is no doubt due to the use of a full cone of illumination instead of a narrow annular cone, as phase contrast objectives with different annulus widths and radii produce haloes of varying extent.

The selection of the appropriate pattern of interference microscope for the work envisaged is a matter on which the manufacturers should be consulted. Quite obviously the user must make himself familiar with the techniques of measurement and manipulation by practical instruction with the microscopes themselves; they are specialized instruments, and cannot usefully be approached in general terms.

The Polarizing Microscope

The formation of an image by the action of a birefringent specimen on polarized light has been described earlier. The process may be considered as a special case of interference microscopy in which both the rays concerned traverse the same geometrical path through the specimen, but undergo a mutual phase shift owing to the difference in their optical path lengths caused by the difference between the refractive indices affecting rays polarized in the two perpendicular transmission planes. Like the interference microscope, the polarizing microscope is principally used to measure the characteristics of the specimen which its special action makes visible, and its design is controlled by this requirement. Mechanical measurements of angular values are made by means of a graduated rotating stage, and can be extended to solid angles by the use of a stage fitted with gimbals, and measurement of thickness is performed with the fine adjustment, which is primarily a micrometer. Measurements of optical path difference are carried out by the use of auxiliary birefringent **compensators**, which can be inserted into the optical system in known orientations, and adjusted to neutralize the effects due to the specimen.

A great deal of work can usefully be carried out without the full range of possible refinements, and polarizing microscopes are, therefore, constructed in varying degrees of complication. In its simplest form, the instrument is required to display the presence of double refraction in the specimen, and to measure the angle between the edge of a crystal and its optic axis in parallel transmitted light. The minimum specification is thus a microscope without a substage condenser, fitted with a graduated rotating stage, a polar filter below the stage, and one at a convenient point above it. The polarizing directions of the two filters in the crossed position are indicated by crosshairs in the eyepiece, which must, therefore, be located so that it cannot be accidentally turned.

One of the two filters should be capable of easy removal from and insertion into the optic system, and one should be capable of rotation, to allow it to be set in the crossed or parallel position with certainty. As a rule it is most convenient to make the substage polar (the **polarizer**) rotatable, and the upper polar (the **analyser**) removeable.

Work is generally carried out at a low magnification, with a 10x objective, which must not itself exhibit any polarization due to strain or other causes. In spite of the low magnification, a graduated fine adjustment is necessary to act as a micrometer. The reliability of the graduations in this case takes precedence over other considerations, as certain types of adjustment are not strictly linear in their action. For this reason the firm of Swift, with an unrivalled experience in these matters, retained a direct-acting micrometer fine adjustment or else a differential screw in their finest instruments until the war.

Until recent advances made possible the use of thin polarizing filters, these simple needs could not conveniently be secured in a microscope with a 160 mm tube, and a certain amount of mechanical complication was necessary. As polarizing microscopes with prisms are still in common use, it is best to describe these rather than the modern simplified ones which are practically biological stands with a rotating stage and polars.

The chief difficulty with polarizing prisms, which exist in several modifications of the original Nicol pattern, is that they are essentially thick and of small free aperture. This makes it necessary to locate them in the optical system where the light rays are contracted—immediately above the objective or in the eyepoint. A prism above the eyepiece occupies the proper position of the eye, and is traversed obliquely by the marginal rays; it is to be avoided when possible. Above the objective it is easily mounted, and is traversed only by rays which are practically parallel. Its thickness, however, results in a change of tube length, and therefore of focus, when it is inserted or withdrawn, and it also produces astigmatism. Where a long tube length is used, these effects are not obtrusive, but with a short tube they must be compensated. The focal difference is easily balanced by using a block of glass in the prism carrier, but astigmatism has to be corrected by a lens, and this changes the magnification as the analyser is introduced and withdrawn.

The most satisfactory method of evading the difficulties is by using **telecentric** objectives, corrected for an infinite tube length, so that the rays travel parallel through the prism, and are then made to converge to the primary image plane by a fixed lens in the microscope tube. This removes the incidental difficulties, and at the same time allows the prism to be set somewhat further up the tube.

With modern polar filters the difficulties do not occur. The filter is unlimited in aperture and effective in oblique light, and can be used as a cap analyser without difficulty.

The substage prism does not cause any difficulty in parallel-light illumination, as its aperture is adequate to cover the specimen, but when convergent light is required, an R.M.S. size of condenser is necessary, as a prism cannot cover one of large size. For crystallographic reasons, there is rarely a need to exceed an aperture of 1.0 N.A., as the condenser is used to provide a cone of inclined rays to explore the crystal directions, and not to generate diffraction spectra for resolution. With a small condenser this implies a short focal length and powerful curves on the lenses, which lead to a certain amount of depolarization, so that the modern use of a polar filter with a full-sized condenser is preferable, though it ought to be a corrected pattern rather than the usual Abbe type.

The condenser must be centrable to the rotating stage. Small types are usually made to swing out of the optic axis, but with large ones an iris diaphragm is used nowadays to avoid having the light convergent when this is not required.

The stage is mounted in a revolving bearing to allow the specimen to be orientated in the plane of polarization. As it is necessary for the specimen to remain centrally in the field of view with different objectives, these are frequently fitted with centring changers, so that all can be aligned with the centre of rotation of the stage. A centring stage is less useful for this purpose unless the centration of the objectives is mutually exact; centring movements applied to the stage, objectives, and condenser also tend to produce a somewhat chaotic arrangement which it is easy to mis-align.

A great deal of work can be done with the simple polarizing microscope designed for observations in parallel light, and it is adequate for most chemical uses, such as comparison of samples fused together on the slide to determine identity or adulteration. Greater applicability is obtained if it is possible to introduce a crystal plate of known retardation between the polars, as this enables the fast and slow transmission directions of the specimen to be identified with respect to its crystalline shape. A slot is, therefore, usually provided below the analyser into which a gypsum plate or a quartz wedge can be slid, to raise or lower the polarization colour of the specimen. It is convenient to be able to see the graduations on a wedge in focus with the specimen, and it is, therefore, often inserted in a slot in the focal plane of the eyepiece, and used with a cap analyser. Compensators are always used with their fast and slow directions located at 45° to the polarization planes, which are always arranged N–S and E–W in the field of view. Before the war, British microscopes had the vibration plane of the polarizer in the N–S direction, but lately the European plan, which is the converse, has been adopted. It is important to know which is which. Mention must be made of the celebrated Dick design in which the stage is fixed, and the polars rotate together with the compensators; this is especially useful with small specimens, as it eliminates centring difficulties.

Chapter Ten

THE ILLUMINATION
OF OPAQUE OBJECTS

Objects which do not transmit light even when prepared as thin sections are extremely scarce, and practically confined to the metals. It is often useful, however, to examine a specimen without preparing it for formal mounting, and there is no doubt that such examinations could be made much more commonly than is usually the case. A solid object such as a leaf or an insect or a piece of coal, when examined by top lighting is seen just as it is in ordinary circumstances—that is, by the distribution of reflection and diffusion which results from the texture of the surface and its superficial transparency. Very often a stereo-microscope is used to increase magnification (Fig. 127). The result is a picture which is at once understandable in familiar terms, and so is especially useful for instructing beginners.

There is little difficulty in these days of electric lamps in providing top illumination of adequate intensity, though formerly the mechanical limitations of flame lamps which could scarcely be directed downwards led to the introduction of a mass of apparatus as inconvenient as it was ingenious. Top lighting does not require great optical accuracy in casual use, and the necessary intensity is quite low. Frequently no more is required than a bench lamp

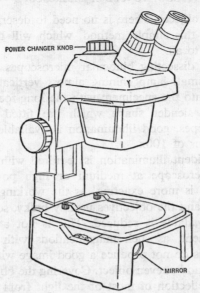

POWER CHANGER KNOB

MIRROR

Fig. 127. Bausch and Lomb stereomicroscope with continuous magnification changer from 0.7x to 3x.

brought close to the microscope, though attention must be paid to the considerable heat radiated from the shade and bulb.

Such lighting, from an opalescent bulb, tends to provide a diffuse illumination, as the source is extensive with respect to the field of view, and this reduces the appearance of relief by minimizing shadows. Should these be of interest, or unilateral lighting be desired for other reasons, nothing is simpler than the use of a flash-lamp bulb of the lens-front type clipped to the microscope stage to shine at the desired angle.

Such a device is readily extemporized, and is entirely adequate for low-power examination of uncovered objects. There is no need to describe variations of this simple method, which will be readily obvious to all.

Many dissecting binocular microscopes are fitted with a small lamp shining almost vertically downwards onto the specimen; with the long-focus objectives of slender shape which are fitted to these microscopes, good illumination is available at magnifications of 100x.

If incident illumination is required with an ordinary microscope at medium or high powers, the situation is more exacting, as the working distance is small and the objectives usually bulky, so that the space available is reduced. This is not a common requirement, as such magnifications with ordinary objectives do not produce a good image with an uncovered and uneven object. Covering the object leads to the reflection of most of the light from the glass surfaces, and introduces severe glare, which obscures the image.

Special apparatus has been designed for such conditions, and consists essentially of a condensing system surrounding the objective and providing a hollow cone of oblique light outside the acceptance angle of the objective. This may be compared with dark-field illumination, as it reveals only features which are rough or inclined to the optic axis, and so deflect the incident light back into the lens; a perfectly smooth and polished specimen would be invisible.

The most venerable of these devices was introduced by Lieberkühn about 1700, and consists of a parabolic metal reflector which slides on the objective body. The specimen, which must be fairly small, is mounted above a black disc, and light from the plane mirror passes up around it to the reflector, which reflects it down to the focal point of the objective. In its original form, as supplied in the nineteenth century, it is quite obsolete, but the Beck aplanatic ring illuminator, consisting of a corrected Mangin mirror, represents a corrected design of the device, and is capable of use with a 40x objective.

Modern apparatus for this purpose uses light from above the stage, and so is applicable to specimens of unlimited extent. In effect, the Lieberkühn reflector is inverted, and used to bring to a focus a hollow beam of light passing downwards around the objective (Fig. 128B, C). This light is provided by a lamp at one side of the nosepiece, which is directed onto a ring-shaped reflector set at 45°. This deflects the illuminating beam downwards as a hollow beam around the image-forming beam passing up from the objective through the central aperture. Individual objectives are mounted in their appropriate condensers, which can be adjusted by a screw thread, and attached to the lamp unit by a dovetailed slide. The original Chapman design, made by R. & J. Beck, Ltd., uses reflecting condensers (Fig. 128C), whilst Messrs. Leitz employ annular lenses in their Ultropak (Fig. 128B).

Such an illuminating unit is of great value both for visual examination and for photography, and a considerable range of powers is available, including both water immersion high powers and attachable fronts to allow a low power to be used on a wet specimen. It is perfectly possible to use the objectives with transmitted light instead of incident lighting, though a mixture of the two is not helpful unless

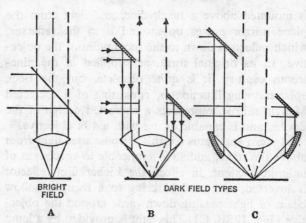

Fig. 128. Incident light systems.

colour contrast between them is used. The only draw-back is the length of the component, which restricts it to a large microscope with plenty of space between the tube and the stage; this provides the necessary focusing action and permits specimens to be supported on a block laid on the base or the bench itself.

Bright field vertical illumination, in which the objective itself acts as condenser, is practically confined to metallography, and is used with uncovered specimens polished and etched in standard ways. In this case, light is introduced laterally into the microscope tube and reflected axially downwards by a thin glass reflector (Fig. 128A). The objective condenses the light onto the specimen, and transmits the image-forming rays to the eyepiece through the transparent reflector. There are considerable possibilities of glare in this procedure, and the introduction of lens-blooming has resulted in great improvements. The reflector is now coated to increase its reflecting power, which produces a brighter image, because more is gained by brighter illumination than is lost by decrease in image transmission, and the objective lenses are coated to decrease reflection and so minimize glare from their surfaces.

It would not be advantageous to embark on a necessarily condensed discussion of metallographic microscopy, as the techniques employed are quite distinct from those used by microscopists generally. Metallurgical microscopes are now generally built as integrated units for their special purpose, and are incapable of other uses. Characteristically they have their coarse focusing action applied to the stage instead of the body, to avoid disturbing the alignment of the illuminating beam when a separate lamp is used. It used to be considered that for one adjustment to move the body and the other the stage was mechanically ill-advised; the probable reason for this dictum was the difficulty of securing parallelism between two hand-made slides separately fixed to the arm. Mechanical construction has robbed this of almost all its significance, and in the case of the metallographic microscope, where the illumination comes from the objective in the first place, the only possible requirement is that the stage surface should be perpendicular to the optic axis. It would not matter much if the stage rackwork were obliquely located.

Chapter Eleven

RECORDING MICROSCOPIC
IMAGES

When a permanent record of a specimen is required
for future reference, or to facilitate discussion, it
may be drawn or photographed. There is no doubt
whatever that for instructional purposes a good drawing of a specimen is far more satisfactory than a
photomicrograph; drawings can vary in status from a
rough free-hand sketch or cartoon, in which the
relevant points can be emphasized and irrelevancies
suppressed, to meticulously exact camera-lucida or
projection drawings suitable for measurement. The
great advantage which they possess is the ability to
show the image as it is seen, with the focal latitude
which results from the ocular accommodation of the
observer, who can also change the focus to demonstrate the relationships of the parts of a thick specimen.

Drawing

A good sketch is, therefore, preferable to a photograph when attention has to be directed to particular
features, as is almost always the case in instructional
work; not only is the specimen better displayed, but
it can be appreciated with less effort and chance of
misinterpretation. This is the unanimous concensus

of opinion among instructors, and the facile view that photography has made drawing unnecessary must be regarded as a mischievous heresy. Even in the case of a solitary student, it will be found that the habit of drawing contributes in no small measure to an understanding of the specimen; the eye does not appreciate instantly the content of a view, and the extended time of observation necessary for drawing combined with the unconscious appreciation of the exact relationships of the parts provides a memory which will be compared with the new image when a similar specimen is encountered in the future.

The simplest method of drawing a specimen is that practised in any elementary laboratory, in which the microscope is placed vertically, and the specimen examined with the left eye whilst the right one looks down on the paper and pencil, which moves round the image. The ability to make sketches in this fashion requires a certain relaxation of the reflex co-ordination; normally the eyes converge as one looks downward, and for this purpose they must be controlled to remain parallel or even become slightly divergent. Initially, therefore, there is some difficulty in maintaining a stable condition, and the image seen tends to wander across the paper, whilst the sketch becomes distorted. With practice, however, very creditable pictures can be produced, and linear proportions scaled from them. This method of drawing is useful on account of its independence of mechanical aids, and once the trick has been acquired will be used in preference to more formal methods. The exasperated beginner may be reassured by the knowledge that very many thousands of other students have previously acquired the knack, and have not suffered permanently from diplopia, watering eyes, and headache.

It is possible to make this process easier than it otherwise is by using the microscope in an inclined position, with the paper attached to a board at a corresponding inclination. In general it is easier to see comfortably when looking obliquely downwards rather than vertically, and there is no real reason for perpetuating an inconvenience which arose in the first case because the European laboratory microscopes lacked an inclination joint, and had to be used upright.

Microscopical drawing was of course practised as far back as the seventeenth century, when Hooke's Micrographia, Leeuwenhoek's letters to the Royal Society, and Malpighi's anatomical studies showed what could be done by expert draughtsmen. It is appropriate to mention with these masters Tuffen West, whose production of the thousands of microscopical drawings used as illustrations in all the major Victorian textbooks was acknowledged by their authors. These careful and exact pictures were a model of microscopical artistry, and can still be usefully consulted.

The use of optical accessories to assist the drawing process certainly dates back to the eighteenth century, and many types of varying complexity have been produced. At the present time these are represented by the **eyepiece reflector,** which can be used to project the image down onto the paper without directly observing it, and the **camera lucida,** which reflects the image of the pencil and paper into the eyepiece.

Camera Lucida

The camera lucida used now is the pattern originated by Abbe, and consists of a beam-splitting prism which deflects part of the line of sight horizontally sideways onto a mirror held on a side arm, which brings the paper into view. The only essential difficulty with this device is that the field of view is restricted, as the prism occupies the true eyepoint. It is usual nowadays to provide light filters or a polarizing intensity control in order to secure a convenient balance of brightness between the images of the specimen and the pencil. In most models of the Abbe camera lucida the mirror is not set directly over the paper, so that the virtual image of the eyepiece diaphragm is elliptical instead of circular; this can be countered by tilting the drawing board towards the microscope, but it is rather uncomfortable to

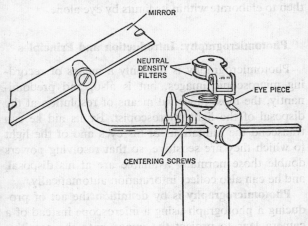

Fig. 129. American Optical camera lucida in place above an ocular.

draw in these circumstances. In some examples (Swift–Ives) the lateral mirror is omitted, and a deflecting prism used which diverts the line of sight from the eyepiece to the paper.

The camera lucida is essentially applied to the monocular microscope, but there is no real reason why a binocular should not be used if necessary. The vast majority of these, however, are made with the right eyepiece adjustable, which prevents the attachment of the camera lucida; Leitz instruments have the left eyepiece adjustable.

The camera lucida should be used only for the general outline of the specimen, and not for minute detail; as has been explained, the draughtsman in his finished picture shows what the accommodation of his eye and also small focal changes reveals. To alter the focus of the microscope during a drawing introduces changes in magnification and usually in position; it is thus better to put the framework down, and then to elaborate within its limits by eye alone.

Photomicrography: Introduction and Principles

Photomicrography is not only a means of recording microscopic images, but is also, and predominantly, the most powerful means of resolution at the disposal of the light microscopist. By its aid he can transcend the limitations of his eyes and of the light to which they are sensitive, so that resolving powers double those normally available are at his disposal, and he can also collect information automatically.

Photomicrography is by definition the act of producing a photograph using a microscope instead of a camera lens to project the image onto the sensitive surface. The production of enlarged images by using camera lenses with extension tubes, or by means of photographic enlargers, is not strictly photomicrography but **photomacrography,** though there is a region where both processes may overlap.

Photomicrography is older than is generally realized. Daguerre demonstrated his photographic process in August 1839, and the Reverend J. B. Reade exhibited a photomicrograph in the same year. The astounding J. B. Dancer demonstrated the entire process of taking a photomicrograph before a large audience in the same year, going on to microphotography, stereophotography and commercial processing at the same time that he was inventing the electric make-and-break, making apparatus for Joule to determine the mechanical equivalent of heat, and investigating air pollution amongst other activities of comparable calibre.

Photomicrography falls naturally into two divisions, the optical and the photographic. The modern tendency is to provide apparatus which is as nearly automatic as possible, and to reduce the microscopical aspect to routine focusing. In this way it is possible to produce a large quantity of miniature film, which can be processed in specialized photographic conditions. The optical difficulties are relegated to the designer of the apparatus, and the sole remaining embarrassment is the filing of the negatives.

Work of this type is essentially of a routine nature, and limited in scope by the facilities provided by the apparatus. Nevertheless it does provide opportunities for original work which cannot be carried out in other ways; the time-lapse photomicrography of tissue cultures, which enable the processes of their growth to be recorded and projected at increased speed, demands a fully integrated apparatus, in which the exposure, film transport, and momentary increase in the level of illumination can be entrusted to a clock system. The main interest of such systems, however, is the mechanical aspects of the design; optically they offer little interest, and their operation requires only a rudimentary microscopical experience.

The use of 35 mm film for photomicrographic purposes has the effect of ousting the photographer also from the process. It is almost invariably the case that 35 mm material is developed in the length; if individual pictures were required, a more convenient negative size would be chosen. Consequently after a suitable type of film and developer has been selected, and an exposure determined, the production of the negative is a matter of purely mechanical routine. It certainly avoids the danger of producing misleading pictures by interfering with the development, but it tends to produce a number of unnecessary negatives taken to ensure that one may show what is required.

This routine recording is the microscopical equivalent of snap-shot photography, and while it is essential in its proper sphere, it cannot be considered as instructive from the photomicrographic point of view; the art must be learned a harder way.

Photomicrography with ordinary apparatus becomes mainly an exercise in projection. The essential requirement is the provision of a good image projected onto a screen; the relationship of the image on the screen and that on the photographic plate is a matter of pure photography, but it is absolutely essential to start with a good projected image, as no photographic expertize can improve a bad one.

The fundamental requirements are as stringent as in good visual microscopy; the objective must be used in its most advantageous conditions as far as correction and illumination are concerned, and these do not admit of any compromise or relaxation. In particular the collimation of the microscope with the illuminating train must be perfect, as a lack of uniformity in the illumination, which the eye will ignore as irrelevant, will be faithfully recorded by the camera. The earlier remarks about the correction of lamp condensers and the situation of the substage diaphragm are particularly applicable to photomicrography.

The principal differences between visual and projective microscopy are concerned with the second stage of magnification in the microscope, which is carried out by the eyepiece. In visual use, the primary image is located in the plane of the eyepiece diaphragm, which circumscribes the field of view, and lies just inside the focal position of the eye lens, which projects a bundle of almost parallel rays of light from each point in the primary image into the eye, which registers an image apparently ten inches in front of it (Fig. 130A).

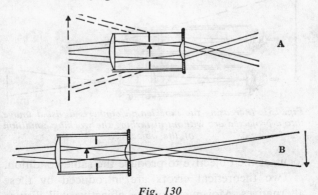

Fig. 130
A. Huyghenian eyepiece in visual use.
B. The same used for projection.

For this image to be received on a screen above
the eyepiece, the relationship between the eye lens
and the eyepiece diaphragm must be slightly changed,
so that the diaphragm is outside the focus of the
eye lens, which will then form a real image of the
diaphragm and the pattern in its plane on a screen
located above it (Fig. 130B). If the distance from
the eyepoint to the screen is the same as that from
the eyepoint to the original visual image, the image
seen by eye and the image projected will be equal
in size; if the projection distance is greater or less
than the visual distance, the projected image will be
amplified or diminished in proportion.

Confining attention for simplicity to a theoretical
object of no thickness, which is either in focus or
out of focus completely as seen visually, it will be
evident that an image can be projected by moving
the entire eyepiece slightly upwards, to bring the
primary image into a plane slightly lower than the
diaphragm (Fig. 131), or by slightly changing the

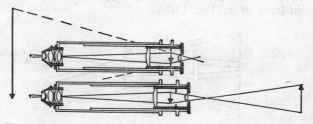

*Fig. 131. Increasing the tube length converts a visual image
into a projected arc without disturbing the working condition
of the objective.*

focus of the objective to produce the same condition.

Two theoretical errors are introduced by these
alternatives. Movement of the objective will disturb

the correction of the primary image, and movement of the eyepiece will alter the tube length and thus the magnification. In any event, the plano-convex eye lens is not at first sight suitable as a projection lens; it might be expected to work better if inverted.

Of these objections, the first is the only one to have any real validity. Except with a very short camera, the movement required to convert a virtual into a projected image is so small that unless the ultimate available resolution is essential it may be disregarded. In common use, the position at which an objective is made to focus will depend to some extent on the eye of the user, and the position of the diaphragm in the eyepiece is not always what the designer prescribed. Nor are eyepieces invariably par-focal, so that the ideal focal position of the objective may not be that which is normally used. Adjusting the focus of the entire microscope by using the fine adjustment, therefore, is commonly acceptable for focusing the projected image.

Where, however, it is worth adjusting the tube length in visual work, so that the objective is actually adjusted to the conditions of observation, it is absurd to project the image by interfering with the position of the objective. In these conditions the change in focus is rightly effected by changing the tube length. Old instruments with a mechanical draw tube are convenient in this respect, but in many modern ones which lack even a sliding draw tube the necessary adjustment can be made by slightly unscrewing the eyepiece fitting. A stiff draw tube may prove worse than useless for this purpose; in these circumstances the eyepiece may be eased up and secured with a rubber band.

The objection to altering the tube length is mistaken; it has the effect not of altering the optical tube length, but of keeping it constant, and the difference in mechanical length produces an insignificant change in magnification.

The use of an ocular designed to produce a virtual image for the projection of a real image is not entirely illogical. It is often overlooked that it is not a case of projecting a flat image by means of a plano-convex lens used in its worst position; the eyepiece should be considered as a whole—as, for example, a Ramsden eyepiece might be considered in such conditions, with both its lenses affecting the result. The image presented to the eyepiece is not a flat image at all, but a curved one, and the effect of projecting this through the eye lens is by no means obvious.

Projection Eyepieces

Special photographic or projection eyepieces have been produced with the object of providing a better projected image by improving the eye lens of the ocular. They usually consist of a Huyghenian eyepiece with a corrected eye lens, often a bi-convex triplet, which is mounted in a threaded tube (Fig. 132). For visual use the eye lens is screwed home,

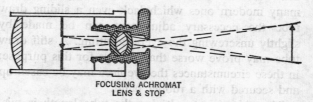

FOCUSING ACHROMAT
LENS & STOP

Fig. 132. Construction of typical projection eyepiece.

and the picture projected by unscrewing it until the eyepiece diaphragm is focused on the screen. The eyepoint is fitted with a diaphragm to remove stray light. These projection eyepieces have a small field diaphragm, to restrict the field of view to an area which is substantially flat, and so lose much of their utility in circumstances where they might be most valuable—the photography of wide fields of view at low powers, where field curvature is serious.

Another projection device for producing a flat field is a negative lens combination, of the type made by Zeiss Jena under the name Homal or by Bausch and Lomb as the Ultraplane. These intercept the rays from the objective before they form the primary image, and transfer their focal position to the screen (Fig. 133). In effect, they turn the microscope into

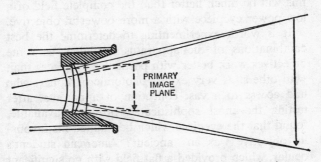

Fig. 133. Negative lens combination for projection.

a kind of Galilean telescope, or the objective into a telephoto lens. As there is no virtual image, they cannot be used for observation; the microscope must first be set up and focused, and the projection system then substituted for the eyepiece and focused on the screen. The Zeiss Homals are designed for use with specific objectives, and cannot be immediately applied to the microscope, as they require adaptors to locate them inside the microscope tube, which must be sufficiently wide to accommodate them.

The especial utility of such devices has been diminished by the introduction of flat-fielded objectives such as the Zeiss Planachromats and the Leitz Plano-objectives, which can be used both visually and for projection.

When ordinary eyepieces are used with ordinary objectives, the curvature of the field which is so much more obvious than in visual conditions can be greatly reduced if the necessary magnification is obtained with the lowest objective power in combination with the highest eyepiece which provides the desired resolution and image size. In these circumstances only the centre of the field of view will be recorded, and this will be much flatter than the complete field of a low power eyepiece with a more powerful objective.

It is worth experimenting to determine the best combinations of the apparatus available, as some objectives work better with particular eyepieces than with others. A very expert photomicrographer, who had access to a vast collection of apparatus, after testing the most sophisticated products available, found that the eyepiece which best suited his favourite objective was an "ancient" American student's ocular, which provided a flat field with no significant distortion. This is certainly a matter of luck; manufacturers design their oculars to work with their own objectives, and the best result is, therefore, to be anticipated from contemporary units of common manufacture, made to achieve a complementary correction. Concepts of what is ideal compromise in objective design alter with time and ability, so that in the absence of a clear critical facility it is well to avoid getting the worst of two concepts. This applies, of course, particularly in the case of apochromatic objectives and compensating oculars.

Choosing the Apparatus

The image projected by the microscope has to be observed to ensure that it is properly focused before it is recorded. The oldest method of doing this is by receiving it on a ground glass screen which will later be replaced by the sensitive plate. The screen can be studied whilst the focus is adjusted until the maximum sharpness has been achieved, and the specimen can be adjusted to ensure that the relevant parts will be recorded. When really fine detail is to be recorded the ground glass does not provide sufficient sharpness, and fine focusing is carried out either on a plane glass with a magnifier, or at clear windows in the ground glass, also with a magnifier. The magnifier is used just as the eyepiece of a microscope is used, to convert the real image into a virtual image for greater convenience; it is focused on cross-lines drawn on the underside of the focusing-screen, where the photographic emulsion will presently lie. There is an advantage in using a lens of short focus for this purpose, as its focal plane is better-defined, but usually the image is so dim that additional magnification must be reduced to a minimum, and a 4x lens or eyepiece is advisable. Theoretically the actual plane in which it is focused is immaterial, as owing to the small angular convergence of the image-forming rays there is considerable focal latitude, but unless the negative will not itself be enlarged photographically it is advisable not to risk any loss of sharpness. When subsequent enlargement or the utmost resolution are required, it is worth using an eyepiece set in a frame instead of the focusing screen, and adjusting the focal plane to the level of the plate.

The exact manner in which this is carried out depends on the type of apparatus employed. At present it is fashionable to use vertical cameras, as these take up less bench space, and are applicable to wet preparations. They have the drawback of being awkward to focus, as the ground glass cannot normally be examined without climbing onto the bench, though this can be avoided in the case of cameras with a reflex focusing screen visible from the side.

Almost any of the small cameras on the market

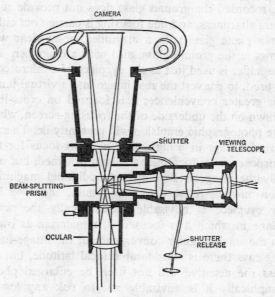

Fig. 134. Leitz "mikas" photomicrographic attachment.

is attachable to any microscope for more or less impromptu photomicrography. Inexpensive adaptors which have draw-tube clamps are available at large photographic supply stores. Reflex cameras with removable front lenses are preferable. In some of the more expensive 35 mm cameras the view finder may be removed and replaced with a magnifying lens system to facilitate focusing. The coarse grain image of the ground glass focusing screen may be improved by placing a cover glass on top of a drop of immersion oil. There are microscope adaptors with their own built-in shutters for cameras with less smooth acting shutters.

With cameras of variable length there is no alternative to inspecting the focal plane, but fixed-length cameras can be fitted with a lateral focusing eyepiece into which part of the light from the microscope is deflected. This side eyepiece is adjusted so that its focus coincides with that of the camera, and enables pictures to be taken without delay. The exact method varies; some makers use a deflecting prism which is withdrawn before the picture is taken, and in some cases the projecting eyepiece is used only for the photographic image, the focusing eyepiece receiving light from below the former. In this case it is possible to use a more powerful ocular in the viewing than in the projection path, allowing a better chance of critical focusing.

Side-viewing is the method of choice for 35 mm photomicrography, where the final image is too small to be studied directly. Nevertheless, the possibility exists of taking a series of pictures out of focus, as a result of faulty focusing. The best method of ensuring sharp negatives is to look at the focusing mark in the eyepiece steadily, and to bring the image onto it; looking at the image and bringing it to the cross-lines is less exact, as the cross-lines are acceptable outside their exact focus to a greater extent than the image of the specimen, owing to ocular accommodation.

In any case final focusing must be performed in light of the colour to be used for the exposure, to avoid possible errors in the chromatic focal planes. This frequently presents difficulty owing to the dimness of an image formed in coloured light compared with normal circumstances. The process is made less difficult if the image is focused with a lens instead of a ground focusing screen, as less light is wasted.

Colour Filters

Colour filters are important in photomicrography. Their principal use is to enhance visibility in the specimen, or to reduce excessive contrast. For this purpose light either complementary in colour to the stain, or else of the same colour, is used. A specimen with blue nuclei and red cytoplasm photographed through a red filter will show black nuclei and little external structure, whilst a blue filter will darken the red portions and relieve the contrast of the nuclei. For contrast, use as complementary filter, and for detail, one of similar colour.

The second purpose of the filter is to match or correct the colour sensitivity of the emulsion. In these days of panchromatic emulsions this is of reduced importance, but with orthochromatic plates the situation existed in which the image focused visually in apple-green or yellow did not register, whilst an out-of-focus image in blue light did so. This was naturally more evident with achromatic objectives than with apochromats, owing to the relative separations of the chromatic focal planes, and the reduction of the light to quasi-monochromatic conditions has the effect of eliminating at least some of the objective errors. The corrective effect of the filter is, therefore, still valuable, though not in its original context.

The third main use of a filter is to enhance resolution by shortening the wavelength of the rays used. As compared with red light, blue light has about twice the resolving power, and by eliminating the longer wavelengths which would only confuse the image, finer detail can be photographed. Carried to its logical end, this process leads to ultra-violet photomicrography. It is essentially a research expedient where the very utmost is required; in other circumstances this effect is of little value. A blue filter is very unpleasant in use, and makes focusing difficult, whilst its advantage over a green screen is comparatively slight.

Choosing the Film, Film Size, and Resolution

Resolution is more readily obtained by photography than by direct observation not only by the use of short non-visual wavelengths, but also because emulsions with resolving powers superior to that of the eye are in general use. It is well-known among photographers that a correlation exists between emulsion speed and grain size, the faster film being coarser and so less capable of recording detail. For photomicrography the most suitable combination is one of ultra-fine grain and ultra-thin emulsion layer, since these produce the sharpest possible image owing to the intrinsic resolving power of the emulsion and the reduction of scatter in the thin layer itself. The slow speed is quite unimportant for still photography, as exposures running into seconds are far more easily governed than those in fractions of a second.

Some of the popular fine grain films for photomicrography now available are: Adox KB 14, Adox KB 17, Adox KB 21, Ilford Pan F, Kodak Panatomic X, Kodak Plus-x Pan, Kodak Plus-x Pan Professional, Perutz Pergrano, Perutz Perpantic, Perutz Peromnia. Development with Fr X-22, Kodak D-76, Kodak Microdol-X generally gives satisfactory results.

Exposure is changed either by changing the shutter speed, by inserting neutral density filters or by varying the intensity of the light source.

The actual size of the negative produced has an effect in terms of resolving power. It may be assumed that a quarter-plate and a 35 mm film frame can be focused to an equal exactitude; the images will require to be of equal size for this purpose, and this will be secured by appropriate focusing-glasses. The larger image will obviously provide better photographic resolution, given identical emulsions, because it is more extended, and in addition it provides an insurance against irregular distribution of the silver halide in the emulsion; a local bare patch in a 35 mm frame is much more serious than the same in a larger negative.

It is perfectly true that 35 mm negatives can be enlarged to an astounding size before grain becomes obtrusive, but in terms of potential resolving power they can never compete with a larger negative. Their advantage lies in the semi-skilled recording of a lot of material where resolution is not a limiting factor. In these circumstances the small negatives

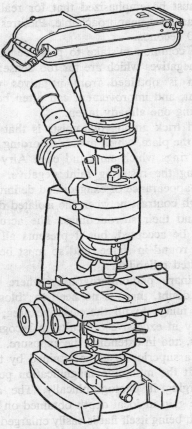

Fig. 135. Polaroid attachment mounted on an American Optical Microstar microscope.

can conveniently be processed and stored in the
length. They cannot be given individual treatment,
and it must be emphasized that for really fast results Polaroid is unapproachable. Pictures are ready
within 10 seconds.

It is a common mistake to produce photomicrographic negatives which are far too dense; the finest
resolution is obtained from negatives which are
quite faint, and improvement can often be obtained
by reducing one already taken.

An old trick among diatomists is that of under-
exposing the plate, which avoids recording light from
the faint rings which surround each Airy disc, and
intensifying the resulting faint negative. This produces an appearance of exaggerated definition owing
to the high contrast between the isolated dots of the
pattern and their surroundings. The actual resolution may be accepted, but it presents all luminous
points as round in shape, and so must be regarded
as of limited reliability.

In ordinary photomicrography where resolution
is not in doubt, there is no need to "blow up" the
projected image to fill the negative area; this may
result only in excessive difficulty in recognizing the
true focus, and in extending the exposure. For many
purposes a superior result is attained by taking the
negative at the minimum magnification permissible,
and enlarging it photographically. The advantage
lies in the gain in focal depth obtained on the negative, which being itself flat is easily enlarged. A flatter
image is obtained by using the minimum objective
power combined with a high eyepiece, rather than a
higher-powered objective and a weak eyepiece; the
field of view is larger in the former case, and only
the central, comparatively flat, portion is of interest.

The minimum magnification permissible in photomicrography is that which ensures adequate separation of the detail resolved by the objective in the case of the emulsion exposed to it. The resolving powers of emulsions are given by their makers; it is advisable to retain a safety factor of about ten as an insurance against irregularity in grain dispersal whenever possible.

Exposure

An exposure meter is a great aid for photomicrography but is not indispensable. Trial exposures can be made under carefully recorded conditions. One must note ocular and objective magnification, N.A., light source details, if variable, filters, film type, exposure time, distance from ocular to ground glass, developer and developing time, etc. The diaphragm of the camera is always kept wide open. Once the correct exposure has been found for a particular set of conditions, the times for subsequent exposures of similar subjects can be calculated by taking into consideration changes in magnification, N.A., emulsion speed, filters, etc. Exposure time varies as the square of the total magnification and inversely as the square of the numerical aperture. If a filter is used it will cut down some of the light so that the exposure will have to be increased by the appropriate filter factor.

Some types of ordinary sensitive photographic exposure meters can be used as aids in determining the correct exposures in photomicrography. Readings are made at the eyepiece or at the focusing ground glass. Any photometer or light meter must be calibrated by photographic exposure tests and a separate calibration is generally required for different equipment setups. Elaborate microscope exposure meters, exposure meters built into microscope-camera adaptors, and automatic cameras are available for many microscopes (see Fig. 14). Synchronous electronic-flash units for high speed photography are also available.

It is appropriate to end a discussion of photomicrography with a note on the relationship between the microscopist and the photographer where both exist in a department. The actual photographic processes in photomicrography only concern the microscopist to a limited degree, and it may be accepted that the photographer will perform them better. The microscopist is well-advised, however, to take his actual photomicrographs himself, as his colleague is by training conditioned to think in terms of images smaller than the object, and so to work by rules which are the reverse of those applying in microscopy. The work may be divided rationally on the basis that the microscopist is concerned with the projection of the image, and the photographer with the selection and processing of the sensitive material, in the light of the microscopist's intentions. This is of especial importance where colour work is to be undertaken, as the relationship of lamp colour-temperature to filter requirements, and the reciprocity corrections which may be required, are peculiarly within the competence of the photographer.

Mention must be made of the instructional leaflets which are issued by photographic manufacturers on their products, and on their application to photomicrography. Whilst these do not represent the sum total of knowledge in this sphere, they do provide sound information on methods which really work;

the user may come to prefer some emulsion which the manufacturer has not considered as likely to suit photomicrography, but he is well-advised to begin with a recommended one, and to develop it in the way suggested.

CONCLUSION

The microscope is at present in a condition of rapid development. It is being applied to a considerable variety of studies not in order to determine their structure by observation, as in the past, but as an adaptor to allow refined physical techniques of measurement to be applied to minute regions of a specimen. This class of application provides the most prolific field for further development; the interest of design today centres less on the optical side than on the mechanical and electrical controlling mechanisms which extend its automatic exploitation in biophysics.

A result of this approach is that microscopes are designed less and less for observation, and increasingly for special purposes; automatic photomicrography, blood-cell counting, and television microscopes with electronic instead of optical control of image contrast are examples of modern developments in which the purely microscopic part is the least important.

It is, however, fallacious to assume that a complex type of instrument must be capable of producing better results than a simpler one of equivalent optical quality properly handled; its virtue lies in its elimination of personal errors, and a good microscopist can often do better work with an instrument which is capable of adjustment to meet variations in working conditions.

There is no substitute for an appreciation on the part of the user of what the microscope is being made to do, and its competence for the work in hand; lacking this, the machine itself is quite capable of providing plausible nonsense, supported by photographs.

An attempt has been made in the preceding pages to provide the basis for a critical approach to instrumental microscopy. It is very far from complete; the argument has been developed as far as possible without recourse to mathematical symbolism, with the intention of giving a coherent general idea of the various factors influencing the integrity of the image. In view of the number of books already existing on the subject it seemed more useful to avoid duplicating what would inevitably be found elsewhere, and to concentrate instead on what is commonly left out, with particular reference to the factors under the immediate control of the microscopist.

It is admitted that in some cases simplification has been carried to extremes, but those to whom this is offensive may find more exact and detailed treatment in books listed in the bibliography. Those who require practical guidance or advice in microscopical matters can always obtain the best available by consulting either the Royal Microscopical Society, the Quekett Microscopical Club, the New York Microscopical Society, or the American Microscopical Society through their Secretaries, and it may be mentioned that nobody need be deterred from this course by diffidence, and no technical enquiry would be rejected. All these Societies are open to microscopists of all degrees of competence and all classes of interest; on the whole the Quekett Club and the New York Microscopical Society represent the amateur experts and the Royal and the American Microscopical Society the professional and research microscopists, but there is a large common membership, and the Journals of these Societies enjoy a very high reputation throughout the world.

GLOSSARY:
OPTICAL TERMS USED IN MICROSCOPY

aberration Any errors that result in image deterioration.

aberration, chromatic A condition in which a lens has different focal lengths for different wavelengths of light.

aberration, spherical A condition in which rays from a point on the axis passing through the outer lens zones are focused closer to the lens than rays passing the central zones.

achromatic Giving a substantially colourless image. Normal correction is for two colours.

annular illumination Radially symmetrical oblique illumination.

aperture The angle between the most divergent rays that can pass through a lens to form an image.

aperture, numerical The product of the refractive index in the object space and the sine of half the angular aperture of an objective.

aplanatic Corrected for spherical aberration and coma.

camera lucida A prism and mirror system which reflects an image of a drawing pencil and paper into the eyepiece.

coma A lens aberration in which a point in the object lying off the axis is imaged as a short comet.

condenser, Abbe A two-lens substage condenser lacking chromatic correction. Modern Abbes are designed for a minimum of spherical aberration and have a very low-angle aplanatic cone.

condenser, dark-field A condenser forming a hollow cone of light with its apex in the plane of the specimen. The image is formed by the diffracted rays only.

coverslips Thin glasses used to cover microscope preparations. The thickness usually varies from 0.1 to 0.2 mm.

diaphragm Any device for controlling aperture. It may be fixed or adjustable.

diffraction When light passes through a small opening or narrow slit, it is scattered or fanned out. The rays are then said to be diffracted.

exit pupil The small disc just above the eyepiece on which all the light leaving the eyepiece converges. It is also known as the Ramsden circle.

eyepiece, compensating An eyepiece corrected for use with apochromatic, fluorite and high power achromatic objectives.

eyepiece, Huyghenian A two-lens eyepiece corrected for use with low power achromatic objectives.

glare Haziness of the image due to the presence of dispersed light.

illumination, critical The formation of an image of the light source in the object field.

illumination, incident Any method of top-lighting a specimen from above the stage.

illumination, Köhler The formation of an image of the light source in the lower focal plane of the condenser or the formation of an image of the field diaphragm in the object field.

illumination, oblique Illuminating the specimen with light inclined at an oblique angle to the optical axis.

illumination, vertical Illuminating the specimen from above with special devices to reflect the light either around or through the objective.

image, real and virtual The image formed by an optical element when the rays, diverging from a point in the object, are converged to a point is a *real image*. If the optical element diverges the rays, they seem to come from a *virtual image*.

infra-red radiation Light of a longer wavelength than the long red rays; invisible to the eye.

interference When light rays interact by reinforcing or suppressing each other, they are said to interfere. Such light is called coherent.

lens A transparent optical element ground and polished to converge or diverge light rays.

magnification The ratio of the length of a line in the image to the length of the same line in the object. Usually referred to as linear magnification.

magnification, empty Magnification beyond that which presents the maximum detail in the image. The usual limit is 1,000 times the working numerical aperture.

mechanical tube length The distance between the shoulder of the objective and the upper end of the body tube.

objective The lens nearest the object. It forms a real, inverted and magnified image.

objective, achromatic Objectives corrected chromatically for two colours and spherically for one colour.

objective, apochromatic Objectives corrected chromatically for three colours and spherically for two colours.

objective, fluorite Objectives corrected chromatically for two colours and spherically for two colours.

ocular *See* eyepiece.

optical axis The line passing through the centers of curvature of the optical surfaces.

optical tube length The distance between the upper focal plane of the objective and the image formed by the objective alone.

polarized light Light in which the vibrations of the individual rays occur in a single plane.

prism A transparent body with at least two polished plane faces inclined with respect to each other.

Ramsden circle *See* exit pupil.

refraction The bending of a ray of light when it enters a medium of different density at an angle.

resolving power The ability of a lens or the eye to render fine detail. The actual resolution obtained may not be as great as the resolving power.

retina The receiving surface in the back part of the inner side of the eyeball.

ultraviolet light Light of a shorter wavelength than the short violet rays; invisible to the eye.

working distance The distance between the front mounting of the objective and the upper surface of the coverslip, when an object with a standard coverslip is focused.

MICROSCOPE MANUFACTURERS AND REPRESENTATIVES FROM WHOM USEFUL LITERATURE IS AVAILABLE

American Optical Company
Instrument Division
Buffalo 15, New York

Bausch & Lomb, Incorporated
84062 Bausch Street
Rochester 2, New York

Ercona Corporation
Scientific Instrument Division
432 Park Avenue South
New York 16, New York (Carl Zeiss, Jena)

William J. Hacker & Company
Box 646
West Caldwell, New Jersey (Reichert, Wien)

E. Leitz, Incorporated
468 Park Avenue South
New York 16, New York

Nikon, Incorporated
Instrument Division
111 Fifth Avenue
New York 3, New York

Tasco Sales Inc.
1075 N.W. 71 Street
Miami 38, Florida

Unitron Instrument Company
66 Needham Street
Newton Highlands 61, Massachusetts

Wild Heerbrug Instruments, Incorporated
Main & Covert Streets
Port Washington, New York

Carl Zeiss, Incorporated
444 Fifth Avenue
New York 18, New York

BIBLIOGRAPHY

Books

Allen, R. M. *Photomicrography*. (2nd ed.), Princeton: D. Van Nostrand Co., Inc., 1958.

Barer, R. *Lecture Notes on the Use of the Microscope*. (2nd ed.) (Blackwell Scientific Publications), Philadelphia: F. A. Davis Co.

Belling, J. *The Use of the Microscope*. New York: McGraw-Hill Book Co., Inc., 1930.

Bragg, W. *Universe of Light*. New York: Dover Publications, Inc., 1959.

Chamot, E. M., and C. W. Mason. *Handbook of Chemical Microscopy* (2 vols). New York: John Wiley & Sons, Inc., 1958.

Clark, G. L. *The Encyclopedia of Microscopy*. New York: Reinhold Publishing Corp., 1961.

Conn, H. J. *Biological Stains*. (7th ed.), Baltimore: Williams and Wilkins Co., 1961.

Eastman Kodak Company. *Photography Through the Microscope*. (3rd ed.), Rochester: 1962.

Françon, M. *Progress in Microscopy*. ("International Series of Monographs on Pure and Applied Biology," V. 9.) New York: Harper & Row, 1961.

Mellors, R. C., ed. *Analytical Cytology*. (2nd ed.), New York: McGraw-Hill Book Co., Inc., 1959.

Needham, G. H. *Practical Use of the Microscope*. Springfield, Illinois: C. C. Thomas Co., 1958.

Schenk, R. and G. Kistler, (trans. by F. Bradley). *An Introduction to the Principles of the Microscope and Its Application to the Practice of Photomicrography*. London: Chapman and Hall, 1962.

Shillaber, C. P. *Photomicrography*. (2nd ed.), New York: John Wiley & Sons, Inc., 1944.

Journals

Barker, R. and R. G. Underwood. "A Cadmium Sulfide Photocell Exposure Meter," *Journal of the Royal Microscopical Society*, V. 80, pt. 1 (1961), pp. 83–87.

Burch, C. R. "Aspheric Reflecting Microscopes and Their Future," *Journal of the Royal Microscopical Society*, V. 80, pt. 2 (1961), pp. 149–161.

Hartshorne, N. H. "Modern Applications of Polarisation Microscopy, Part II," *Science Progress*, V. L, 197: 11–46, 1962.

Haselman, H. "Color Photomicrography—Experience and Practical Application," *Journal of the Royal Microscopical Society*, V. 79 (1961), pp. 277–286.

Huff, C. G. and J. F. Bronson. "Simple Condensers for Ribbon-filament and Mercury Vapor Microscope Lamps," *Journal of the Biological Photographic Association*, V. 24, Nos. 3 & 4 (1956), pp. 121–123.

July, G. "Symposium on Photomicrography III: Choice of Material for 35 mm Photomicrography," *Journal of the Quekett Microscopical Club*, Series 4, V. V (1961), pp. 412–417.

Klosevych, S. "Photomicrography, the Phase Contrast Procedure," *Journal of the Biological Photographic Association*, V. 28 (1960), pp. 89–95.

Lee, J. J. and B. Friedman, "An Electronic Flash Adapter for Photomicrography," *Journal of the Biological Photographic Association*, V. 29 (1961), pp. 93–97.

Montgomery, B. and L. L. Hundley. "A Flying-spot Interference Television Microscope," *Nature*, V. 192 (1961), pp. 1059–1060.

Needham, G. H. "Hints on 35 mm Colour Photomicrography, Especially with the Leica Camera," *Journal of the Royal Microscopical Society*, V. 80 (1962), pp. 235–242.

Norris, K. P. "Some Observations on Microscope Cover-glasses," *Journal of the Royal Microscopical Society*, V. 79 (1961), pp. 287–298.

Pybus, W. "Adaptation of the Vinten 16 mm Scientific Mk. I Camera for Time Lapse Cine-Photomicrography Using an Electromechanical Timer," *Journal of the Royal Microscopical Society*, V. 72, pt. 4 (1961), pp. 369–395.

Pybus, W., Roberts, D. C., and D. J. Trevan. "A Self-contained, Transportable Cinemicroscope," *Journal of the Royal Microscopical Society*, V. 80, pt. 1 (1961), pp. 89–95.

Young, M. R. "Principles and Techniques of Fluorescence Microscopy," *Quarterly Journal of Microscopical Science*, V. 102 (1961), pp. 419–450.

The following journals regularly publish articles of interest to microscopists:

Journal of the American Microscopical Society
Journal of the Biological Photographic Association
Journal of the Quekett Microscopical Club
Journal of the Royal Microscopical Society
The Microscope and Crystal Front
The Quarterly Journal of Microscopical Science
Yearbook of the New York Microscopical Society
Zeitschrift fur Wissenschaftliche Mikroskopie

INDEX

American Museum Science Books are a series of paperback books in the life and earth sciences published for The American Museum of Natural History by the Natural History Press, a division of Doubleday & Company, Inc.

*Asimov, Isaac *A Short History of Biology* B6
*Bennett, Wendell and Junius Bird *Andean Culture History* B9
*Bohannan, Paul *Africa and Africans* B8
*Branley, Franklyn M. *Exploration of the Moon* B1
 Drucker, Philip *Indians of the Northwest Coast* B3
 Hartley, W. G. *How to Use a Microscope* B10
*Lanyon, Wesley E. *Biology of Birds* B2
 Linton, David *Photographing Nature* B7
 Lowie, Robert *Indians of the Plains* B4
 Oliver, Douglas *Invitation to Anthropology* B5

*Also available in a Natural History Press hardcover edition.